CLARK PUBLIC LIBRARY
3 9502 00128 6343

T4-ATW-286

THE STOMACH

AND DIGESTIVE SYSTEM

YOUR BODY YOUR HEALTH

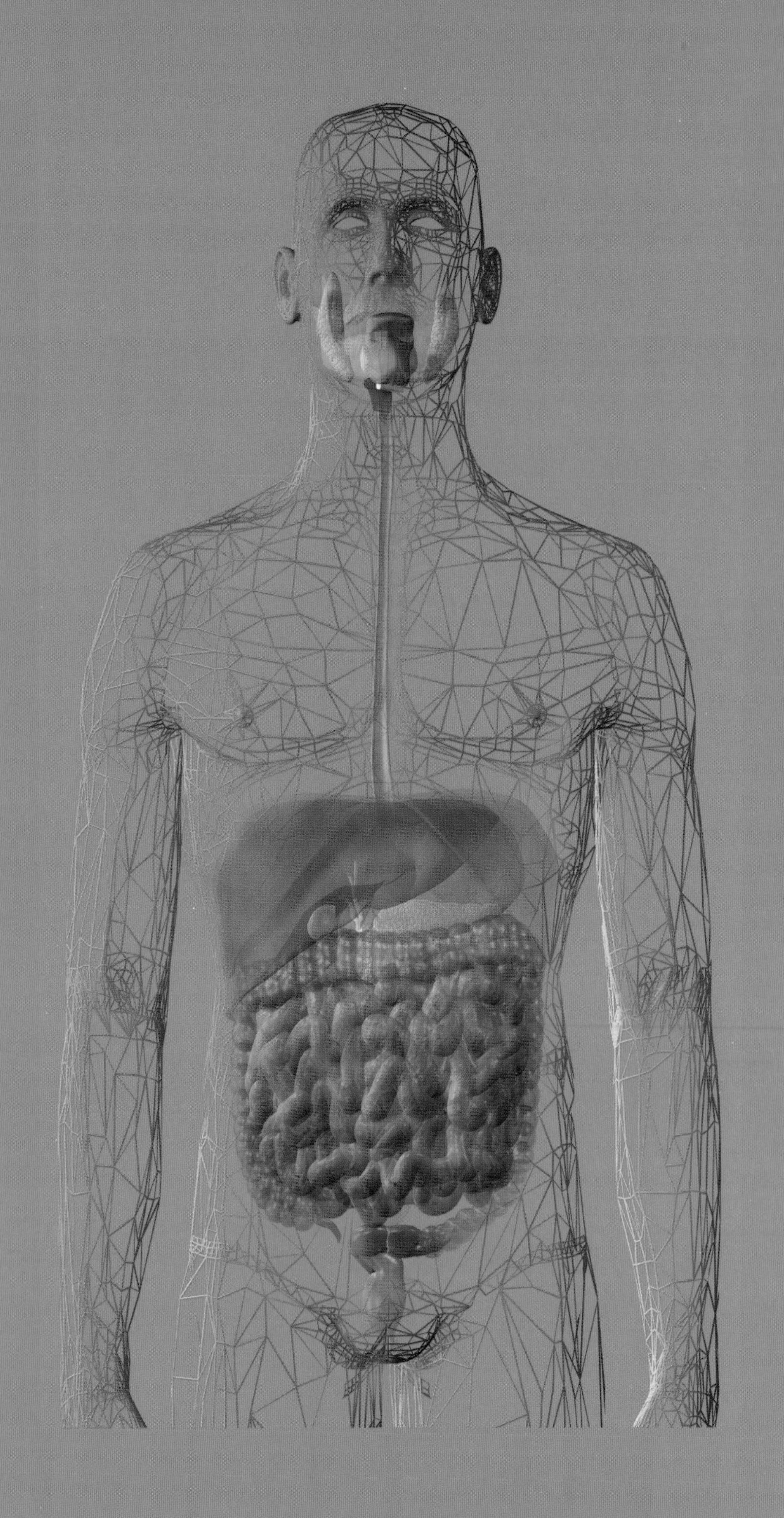

YOUR BODY YOUR HEALTH

THE STOMACH AND DIGESTIVE SYSTEM

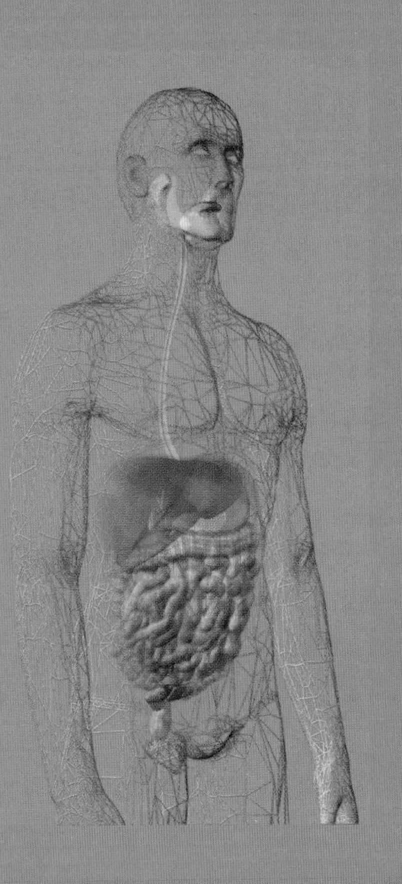

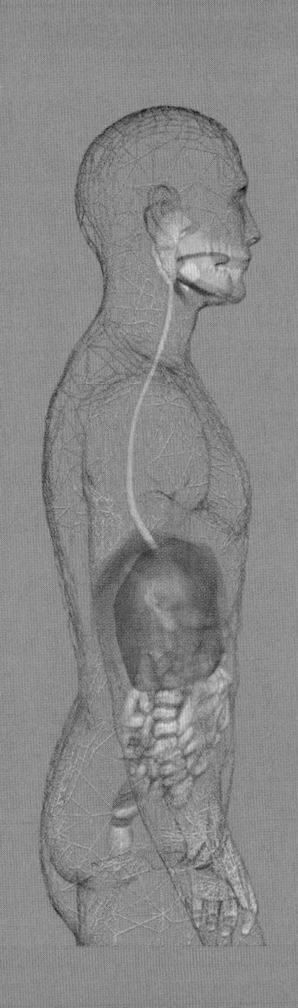

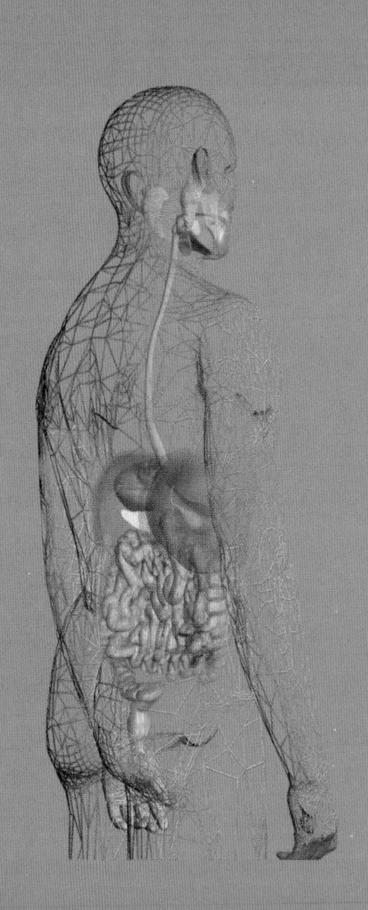

The Reader's Digest Association, Inc.

Pleasantville, New York

London Sydney Montreal

The Stomach and Digestive System

was created and produced by
Carroll & Brown Limited
20 Lonsdale Road
London NW6 6RD

Library of Congress Cataloging-in-Publication Data has been applied for.

ISBN 0-7621-0453-8

Printed in the United States of America
1 3 5 7 9 8 6 4 2

The information in this book is for reference only; it is not intended as a substitute for a doctor's diagnosis and care. The editors urge anyone with continuing medical problems or symptoms to consult a doctor.

IE 0004/IC

American Edition Produced by
NOVA Graphic Services, Inc.
501 Office Center Drive, Suite 190
Ft. Washington, PA 19034 USA
(215)542-3900
http://www.novagrafics.com

President
David Davenport

Editorial Director
Robin C. Bonner

Editorial Assistant
Linnea Hermanson

Composition Manager
Steve Magnin

Art Director
Paul Fry

Gastroenterology Specialist Consultant
Dr Simona Rossi, MD
Thomas Jefferson University Hospital
Department of Gastroenterology and Hepatology

CONTRIBUTORS

Dr Elspeth M Alstead

Dr Saul Berkowitz, MRCP, Department of Gastroenterology, Whittington Hospital, London

Dr Harriet Gordon, MB BS, MRCP, MD, Department of Gastroenterology, Whittington Hospital, London

Delilah Hassanally, MSc, MB BS, FRCS, Department of Surgery, Whittington Hospital, London

Fiona Hunter, BSc, Dip, Dietitian

Joel Levy, BSc, MA, Medical Writer

Dr Geoff Smith, BSc, MB BS, MRCP, Digestive Diseases Research Centre, St. Bartholomews, London

Karan Thomas, BSc, Physical Activity and Health Specialist

Dr Voi Shim Wong, BSc, MB, ChB, MRCP, MD, Department of Gastroenterology, Whittington Hospital, London

For the Reader's Digest
Editor in Chief Neil E. Wertheimer
Editorial Director Christopher Cavanaugh
Senior Designer Judith Carmel
Production Technology Manager Douglas A. Croll
Manufacturing Manager John L. Cassidy

YOUR BODY YOUR HEALTH

The Stomach and Digestive System

Awareness of health issues and expectations of medicine are greater today than ever before. A long and healthy life has come to be looked on as not so much a matter of luck but as almost a right. However, as our knowledge of health and the causes of disease has grown, it has become increasingly clear that health is something that we can all influence, for better or worse, through choices we make in our lives. *Your Body Your Health* is designed to help you make the right choices to make the most of your health potential. Each volume in the series focuses on a different physiological system of the body, explaining what it does and how it works. There is a wealth of advice and health tips on diet, exercise, and lifestyle factors, as well as the health checks you can expect throughout life. You will find out what can go wrong and what can be done about it, and learn from people's real-life experiences of diagnosis and treatment. Finally, there is a detailed A to Z index of the major conditions that can affect the system. The series builds into a complete user's manual for the care and maintenance of the entire body.

In this volume we look at the pipeline and refinery behind your body's energy supply—the stomach and digestive system. Explore why the proper functioning of the digestive system is at the very center of your everyday health and well-being. Discover the amazing subtleties of your ability to taste, the role performed by the vast numbers of hungry bacteria lining your digestive tract, and the astonishing strength of the hydrochloric acid bath otherwise known as your stomach. Why does an aperitif seem to stimulate the appetite, whereas smoking suppresses it? Learn what you can do to safeguard your finely tuned digestive system, why fiber has such profound effects on the functioning and efficiency of the gastrointestinal (GI) system, and which are the superfoods that can give you a real appetite for life. We meet the digestive system experts and reveal how they find out what is wrong, from physical examinations to sophisticated imaging techniques, as well as the wide range of treatments available, from drug therapy to advanced "keyhole" surgery.

Contents

1 How your digestion works

2 Healthy digestion for life

3 What happens when things go wrong

The life story of the stomach

Developing a relationship with your internal organs is difficult. It's hard to feel personally involved with your liver or warm toward your kidneys. Your digestive system, however, can seem almost human: It can gurgle contentedly, rumble plaintively, or churn angrily. The emotional nature of the system is celebrated by common terms like "gut feelings" or "butterflies in the stomach."

Writers and psychologists alike refer to the basic passions of fear, anger, and joy as visceral emotions, meaning literally, the "emotions of the gut." But how did we get so personal about what is essentially a 26-foot-long tube running from the mouth to the anus, encompassing the stomach and the small and large intestines on the way? What is so special about the digestive system?

GUT INSTINCT

First and foremost, hunger is one of the fundamental human drives. The urge to eat is hard-wired into the brain and overrides more intellectual concerns with ease. In fact, the spheres of the brain that generate feelings of hunger and appetite are closely linked to the areas responsible for thirst, sex drive, and the impulse to violence or terror (the fight-or-flight response). This means that the functioning of the digestive system is bound up with a number of primitive desires and drives.

Second, the production, preparation, and consumption of food is a central feature of all human cultures. Quite apart from the sensual enjoyment we gain from eating a delicious meal, the rituals that surround the act have far-reaching social consequences, not the least of which are family bonding and other personal relationships. In other words, the operation of your digestive system has a significance that reaches far beyond the purely biological.

Furthermore, the biological properties of the digestive system are so remarkable that it is a lot more special than you probably imagine. In the complex collaborative process of your body, the digestive system is both supply pipeline and refinery. It channels raw material (food) from

Millions of glands in your stomach produce highly corrosive hydrochloric acid that is so powerful that it can burn skin and dissolve razor blades.

the source (the mouth) through a series of stages of intense processing as fierce as any chemical factory (in the stomach and upper parts of the small intestine) and then extracts the nutrition with an efficiency unparalleled by human engineering (in the small intestine), leaving only waste that is neatly packaged for disposal (in the large intestine) and piped away.

INSTRUMENTS OF MASS DIGESTION

Each step in this vital process involves amazing biological tools. In your stomach, for instance, millions of glands produce highly corrosive hydrochloric acid, which is 1,600 times more acidic than vinegar—strong stuff, but your digestive system can handle it. The stomach lining is shielded by a layer of mucus secreted by cells in its wall, and when the stomach contents reach your small intestine, they are bathed in a flood of bicarbonate, which neutralizes the otherwise lethal acid.

Also inhabiting your digestive fluids are billions of enzymes—special molecules produced by your cells, that act like tiny machines. These nanoscale biomachines go to work on the molecules that make up food—for instance, the starch in a potato or the protein in an egg—chopping them up into smaller bits. Scientists are envious of the amazing processing power of your body's enzymes and occasionally use their secrets in factory and laboratory processes. Many biological laundry detergents, for instance, use enzymes to "eat" the dirt off clothes.

Many of these bacteria are deadly killers if they stray from their usual homes. In your large intestine, for example, the most common bacterial resident is *Escherichia coli*, certain strains of which are the bane of unhygienic food establishments and people with reduced resistance to infection. Not all bacteria deserve such bad reputations, however; many of the species living throughout your digestive system have a vital role to play, undertaking some digestive tasks that your own body cannot perform.

Digesting the dirt
The amazing abilities of your digestive enzymes, such as the pepsin shown at left in crystallized form, are mimicked in biological laundry detergents in order to break down dirt.

TUBE JOURNEY

It should already be obvious that there is a great deal more to your gastrointestinal (GI) system than a rumbling stomach or painful gas. Your digestive tract is part of an extensive and convoluted system that includes the structures of the mouth and accessory organs such as the liver and pancreas. How did so much tissue get packed into your body?

The story begins in the womb, where just three weeks after conception, the tiny embryo has already organized itself into a number of distinct tissue layers. One of these is

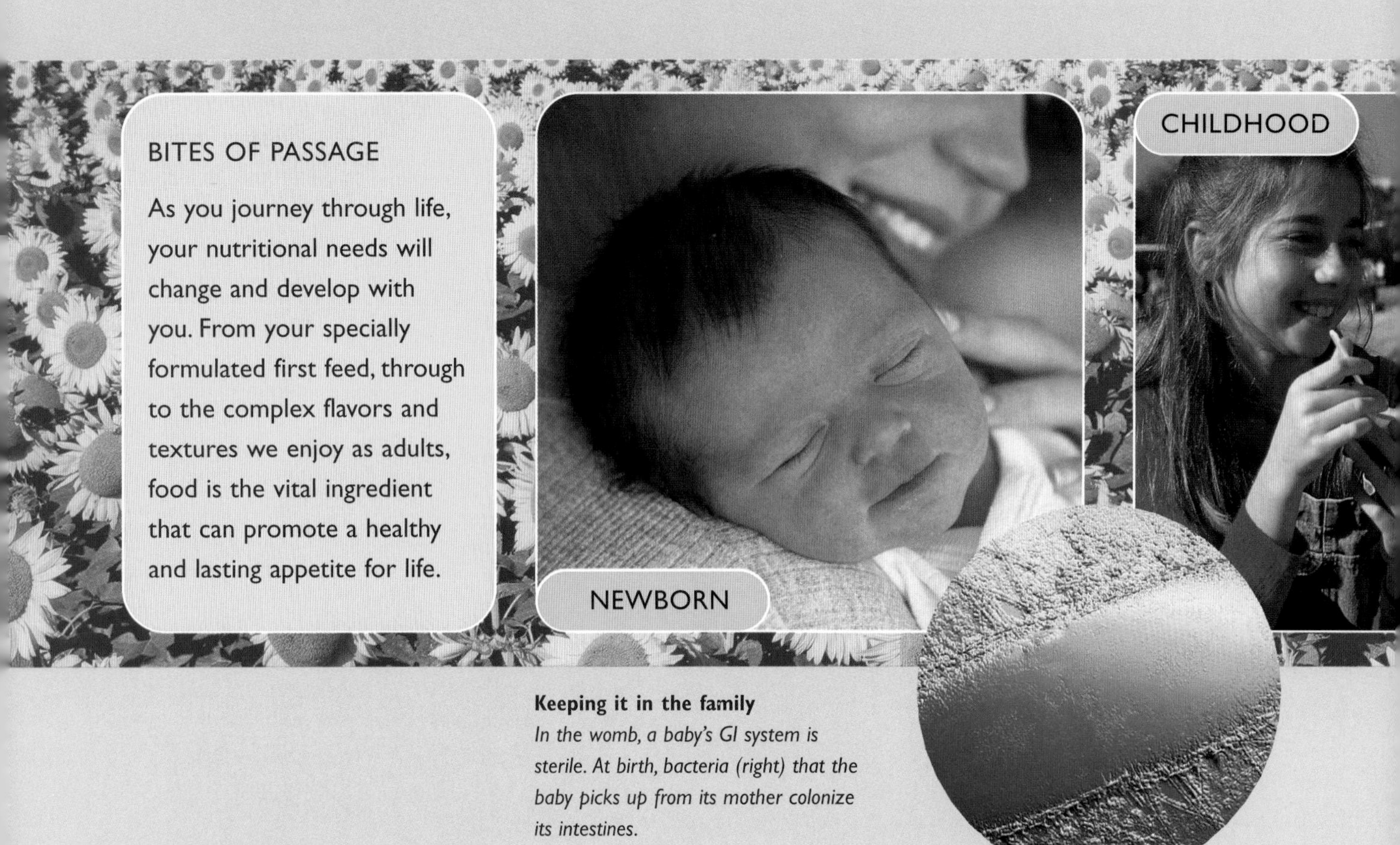

Keeping it in the family
In the womb, a baby's GI system is sterile. At birth, bacteria (right) that the baby picks up from its mother colonize its intestines.

known as endoderm, and the tissues and organs that arise from this layer will form the digestive system. Along the front of the embryo, the endoderm curls into a tube—the embryonic GI system—and on either side of this tube, two cavities appear. These grow and join to give the abdominal space in which your intestines will sit. Meanwhile, the primitive system is suspended between the front and back of the embryo by two sheets of tissue called mesenteries.

By week 6, two pockets of tissue have grown off the sides of the digestive tube and embedded themselves in the mesenteries. These will become the liver and pancreas. Simultaneously, the tube itself is getting longer, outgrowing the tiny space inside the embryo as it turns into intestine. It pushes out into the stalk of what will become the umbilical cord, coiling as it does so. Next to the liver, the tube has expanded to form the stomach. By week 10, the stomach, liver, and pancreas are all in position and the coiled intestines start to move back into the body of what is now called the fetus. The digestive tract stretches all the way from the mouth to the anus of the fetus and can now begin to function in a rudimentary fashion.

In week 11, tiny movements start to pulse along the small intestine. Just over halfway through pregnancy, the baby starts to swallow amniotic fluid from the womb—up to 18 fluid ounces daily—and process it through the developing digestive system. The nutritional value of the fluid is negligible, but experts think that simply going through the motions of digestion helps to stimulate the further development of the system. Therefore, although the baby's main source of nutrition in the womb is the

The energy of youth
Kids need a high-energy diet with proportionally less fiber, so eating starchy white instead of whole-grain bread is actually often better for them.

Keep your GI system in good shape
The efficient functioning of your intestines is directly linked with your overall well-being.

Healthy habits for life
Teenagers have little time to devote to food preparation, but it is vital they adopt a healthy diet to give them the energy they need and to protect them from problems in adult life.

A ripe old age
Looking after your body does not mean cutting out the enjoyment of food. Older adults need far less food than they did, but they should aim to eat more fiber.

umbilical cord, when the time comes for the baby to be born, the digestive system is prepared to receive food via the mouth. Indeed, the newborn's stomach and intestines already contain the necessary acid and enzymes.

MOTHER'S MILK

As mammals, human babies feed on milk for the first few months of life. Ideally, the milk comes from the mother, but this is where human culture starts to get involved with biology. Different cultural groups have widely differing rules about how long a baby nurses at the breast, when to introduce solid meals, and, of course, the type of food given as the child grows into adulthood. Interestingly, cultural practices may have actually caused genetic changes in some populations around the world. Caucasians traditionally consume large quantities of dairy products in their diets, even after weaning, and most adults will retain the special enzyme (lactase) that they need to break down the sugar found in milk (lactose). There are many other ethnic groups, including Indians and Asians, however, who do not generally drink milk after infancy, and consequently they can very often lose the capacity to make lactase after they have been weaned.

The ability to tolerate lactose is just one example of how digestive systems vary among individuals. Your unique genetic and cultural inheritance means that your digestive system is unlike anyone else's. Its response to different foods and drinks is particular to you—for instance, you may be able to eat very spicy food without getting an upset stomach, but bran flakes or brussels sprouts may

All along your digestive pathway, but most notably in your large intestine, lurks a secret population of alien invaders—bacteria, more than 750 trillion of them.

cause you all sorts of digestive problems.

GI HOUSEKEEPING

While each of us is unique, certain factors affect everyone's digestive system. As you age, your senses of taste and smell slowly decline. The muscles that keep food moving through your GI system lose tone, and your stomach and intestines gradually lose efficiency. These changes can make you more vulnerable to disease and other problems such as ulcers or constipation. Apart from a handful of conditions that can go unnoticed, many digestive illnesses make themselves felt immediately with nagging symptoms—from stomachache to constipation. Although digestive problems are not always serious, the digestive system does seem to have a disproportionate influence on your overall sense of well-being.

In fact, the GI system can be called the key to good health. A fit and fully functional digestive system promotes a positive sense of well-being. It supplies nutrients that keep the rest of your body in top form. Enthusiasm for the digestive system as the instrument of healthy living reached a peak in the 19th century, when dietary zealots preached that good digestion was the cure for all ills. Sylvester Graham touted whole-grain bread as a modern-day panacea, and John Harvey Kellogg warned that indigestion caused every disease from typhoid to "mental derangement."

GOOD HEALTH IS GI HEALTH

Although these dietary radicals were perhaps a little over-zealous, taking good care of your digestive system does undoubtedly pay big dividends for your general health. This need not involve a diet of prune juice and Metamucil. Common-sense steps are all you need: Don't eat to excess, maintain a balanced diet rich in fiber, drink alcohol only in moderation, exercise regularly, and practice good food hygiene. The benefits of such simple lifestyle rules include extra energy and vitality; easier and more regular bowel movements; the means to reach and maintain a sensible weight; better-looking skin, hair, and nails; and improved all-round health.

Unfortunately, few of us treat our GI systems with the care they deserve, and illness related to them is one of the fastest growing health problems in the developed world. Poor diet, pollution, and stress are making digestive problems such as irritable bowel syndrome increasingly common. Outside the Western world, the problem is even

The best medicine
A balanced diet that includes plenty of water and fresh produce will help promote good GI health.

Foods of the future
In years to come, we may eat "superfoods" designed to promote particular aspects of GI health and general well-being.

worse. Digestive illnesses are the globe's biggest killers—diarrhea and dysentery killed more than 3 million people in 1998 alone. Medical science and epidemiology have made advances toward solving some of these problems. Notable successes in the developed world include the virtual eradication of cholera, together with a radical reduction of serious diarrheal disease, mainly caused by improvements in sewer systems and water hygiene. More recently, the identification of the *Helicobacter pylori* bacterium as the agent responsible for stomach and intestinal ulcers has opened up new avenues of treatment, and advanced diagnostic technologies are enabling doctors to cut colon cancer mortality rates.

WHAT DOES THE FUTURE HOLD?

In the future, genetic engineering promises to go beyond the battle against disease and to transform the very fabric of our digestive systems. Most current genetic research is performed on bacteria, particularly the common human GI bacterium *E. coli*. The enormous population of bacteria that live in your intestines is an obvious target for genetic engineers. These industrious organisms are already helping your digestive system with tasks it cannot perform by itself. A relatively simple bit of genetic tinkering could transform your GI flora into a trillion microscopic factories, producing nutrients and pharmaceuticals from within. For instance, suppose that some of the bacteria that live in your small intestine could be encouraged to make vitamin C. You would then be guaranteed a lifetime supply of this important health-giving, anticancer nutrient. Fed up with excess gas? Simply engineer bacteria in your large intestine to produce enzymes that can break down the gas into a form that is easily absorbed.

The next step would be to engineer the digestive system itself, to create a genetically enhanced "superhuman." By adding the genes to produce new enzymes, your digestion could become more efficient. By altering the genes that control processes such as fat breakdown and absorption, genetic engineers could reduce the number of Calories you absorb and even limit the amount of cholesterol working its way into your bloodstream, combating cardiovascular disease. Improbable

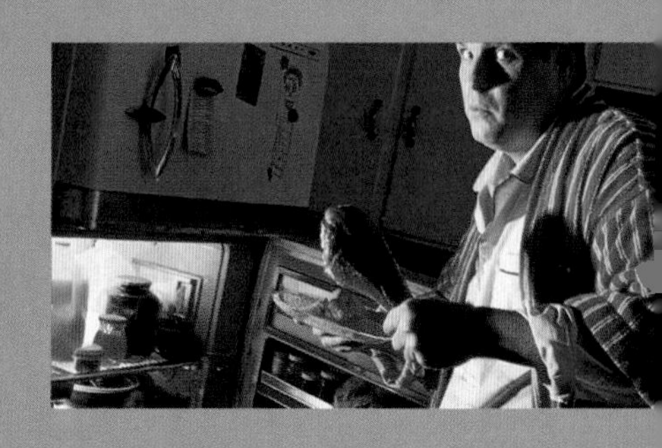

1 How your digestion works

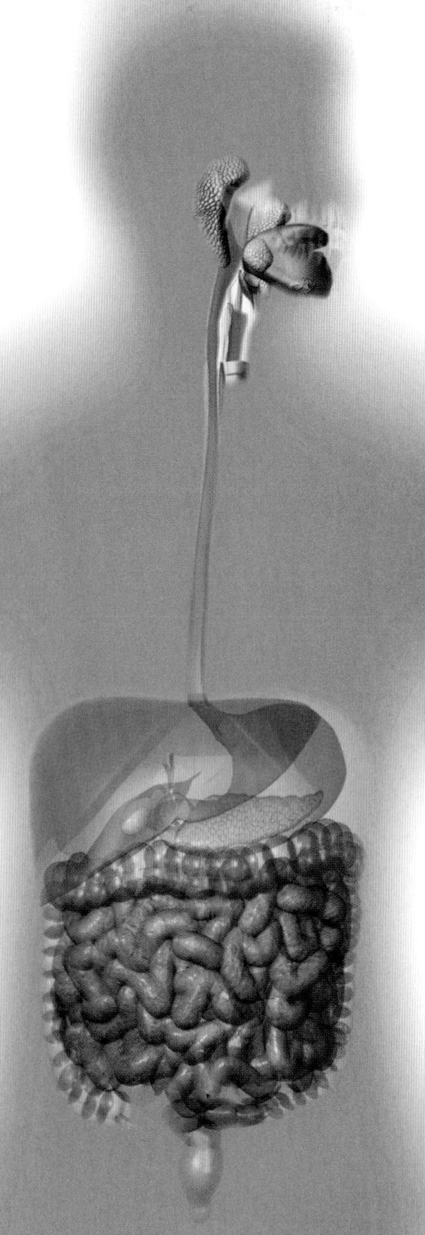

Your amazing stomach and digestive system

Your digestive system is your body's equivalent of an oil refinery. It processes the raw materials of food to produce the fuel that powers every cell in your body and keeps all your other systems running.

LINES OF SUPPLY

Your gastrointestinal (GI) system is essentially a tube leading from your mouth to your anus. This tube has several names, including alimentary canal, gastrointestinal tract, and digestive tract, and is roughly 26 feet long. It is divided into a number of sections: At the top end are the mouth, throat, and esophagus, which leads to the stomach. The stomach feeds into the small intestine, which in turn feeds into the large intestine, which ends at the anus. In addition, three other organs—the liver, pancreas, and gallbladder—supply fluids vital to the digestive process.

The good, the bad, and the ugly

The digestive system has three basic functions. Its primary purpose is digestion—to break down food into its constituent nutrients and to absorb these nutrients into the bloodstream. Second, it has a defensive role, protecting your body by destroying germs or creating conditions too harsh for them to survive. Last, your digestive system disposes of waste, getting rid of not only the indigestible parts of food, but also toxic substances and the waste products of normal bodily processes, such as breakdown products of old blood, by excreting them in stools.

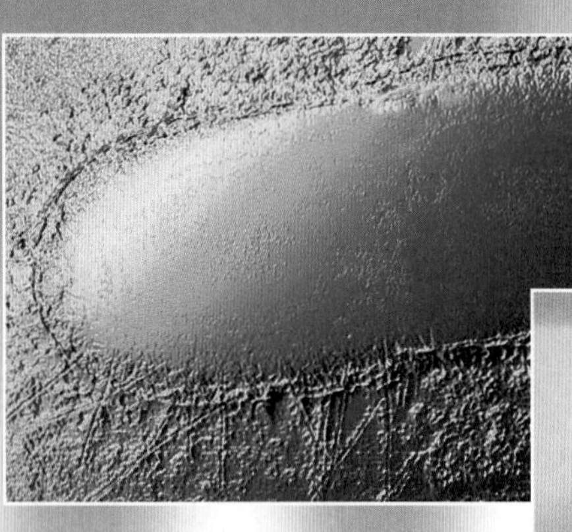

▲ Your friendly digestive flora
Bacteria are normally considered harmful, so it may be a surprise to learn that your large intestine could not function without them; see pages 32–33.

▼ Biochemistry basics
Discover how special proteins called enzymes break down food into its component molecules so they can be absorbed; see pages 30–31.

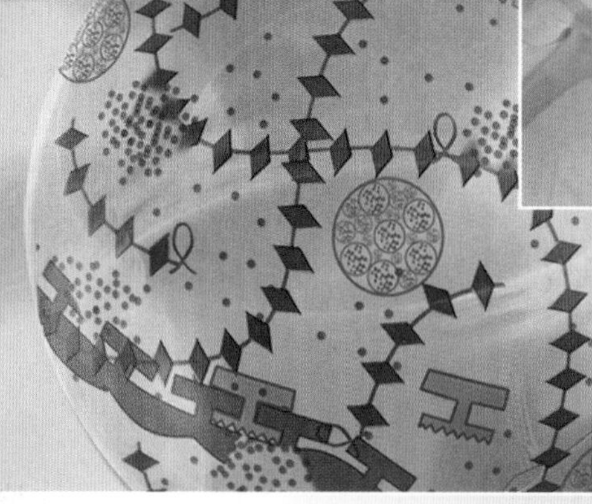

▲ 24-hour control
Each part of the digestive system helps to control the activity of the others. On pages 34–35 we follow the journey of one meal for a day to see how they all work together.

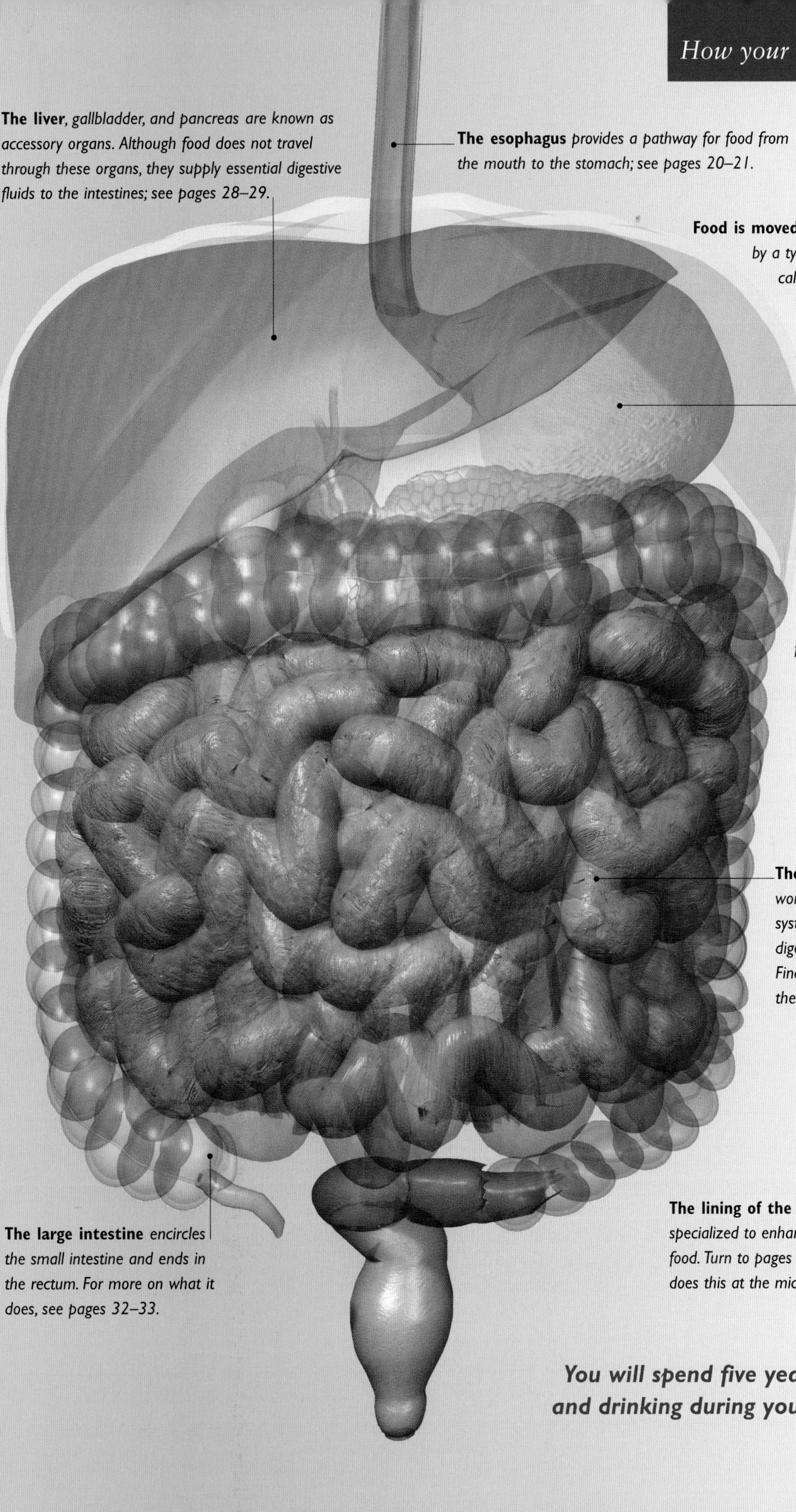

The liver, *gallbladder, and pancreas are known as accessory organs. Although food does not travel through these organs, they supply essential digestive fluids to the intestines; see pages 28–29.*

The esophagus *provides a pathway for food from the mouth to the stomach; see pages 20–21.*

Food is moved along *the digestive tract by a type of muscular activity called peristalsis, described on pages 20–21.*

The stomach
Gastric glands in the stomach produce about $2\frac{2}{3}$ pints of gastric juices every day. These fluids are vital to the stomach's churning and food-processing work; for more information, see pages 22–23.

The small intestine *is the workhorse of the digestive system: 80 percent of digestion takes place here. Find out what it looks like on the inside on pages 24–25.*

The large intestine *encircles the small intestine and ends in the rectum. For more on what it does, see pages 32–33.*

The lining of the small intestine *is specialized to enhance its ability to absorb food. Turn to pages 26–27 to find out how it does this at the microscopic level.*

You will spend five years eating and drinking during your lifetime.

The mouth

As well as being the gateway to your GI system, the mouth is where the first steps in food breakdown occur. Your teeth, tongue, and salivary glands combine efforts to turn solid food into digestible mush.

MAKING AN ENTRANCE

The mouth is more than simply the entrance to the GI system—vital steps in the process of digestion are performed here. The teeth break up food into manageable lumps, ready to be mixed with saliva to produce a semiliquid pulp. Saliva also lubricates the movement of this pulp down the throat and flushes away oral bacteria, helping to prevent dental cavities and gum damage.

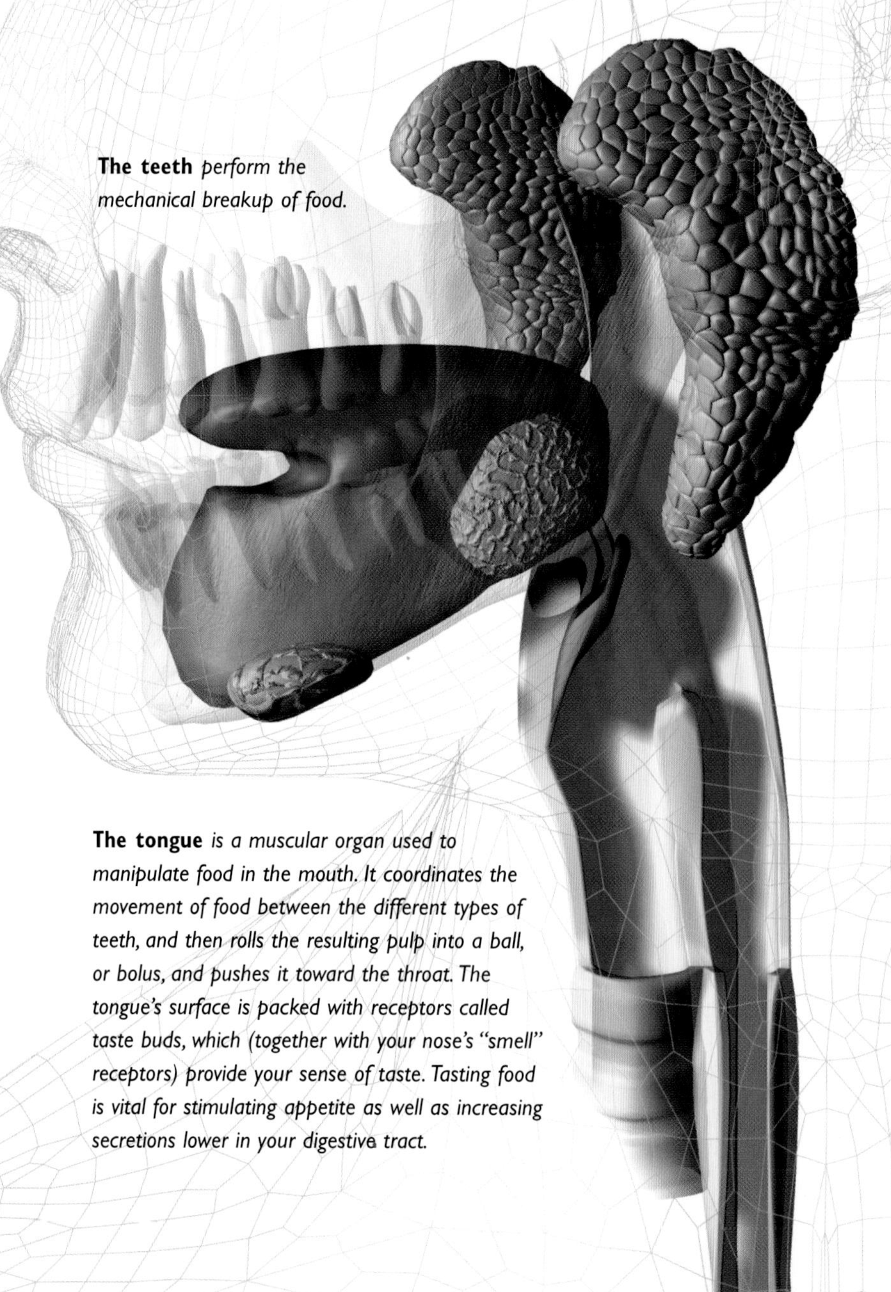

The salivary glands *exist in three pairs—one in the floor of your mouth, underneath your tongue; another closer to the back of the jaw; and the third just in front of your ears. Small connecting tubes deliver $1\frac{3}{4}$–$2\frac{2}{3}$ pints of saliva to the mouth every day.*

The teeth *perform the mechanical breakup of food.*

The palate *forms the roof of your mouth. At the front, it is bony and immobile (the hard palate), whereas at the back, it is soft and flexible (the soft palate). The palate prevents food from going up your nose rather than down your throat.*

In addition to taste buds, your tongue has receptors for pressure, pain, and temperature.

The tongue *is a muscular organ used to manipulate food in the mouth. It coordinates the movement of food between the different types of teeth, and then rolls the resulting pulp into a ball, or bolus, and pushes it toward the throat. The tongue's surface is packed with receptors called taste buds, which (together with your nose's "smell" receptors) provide your sense of taste. Tasting food is vital for stimulating appetite as well as increasing secretions lower in your digestive tract.*

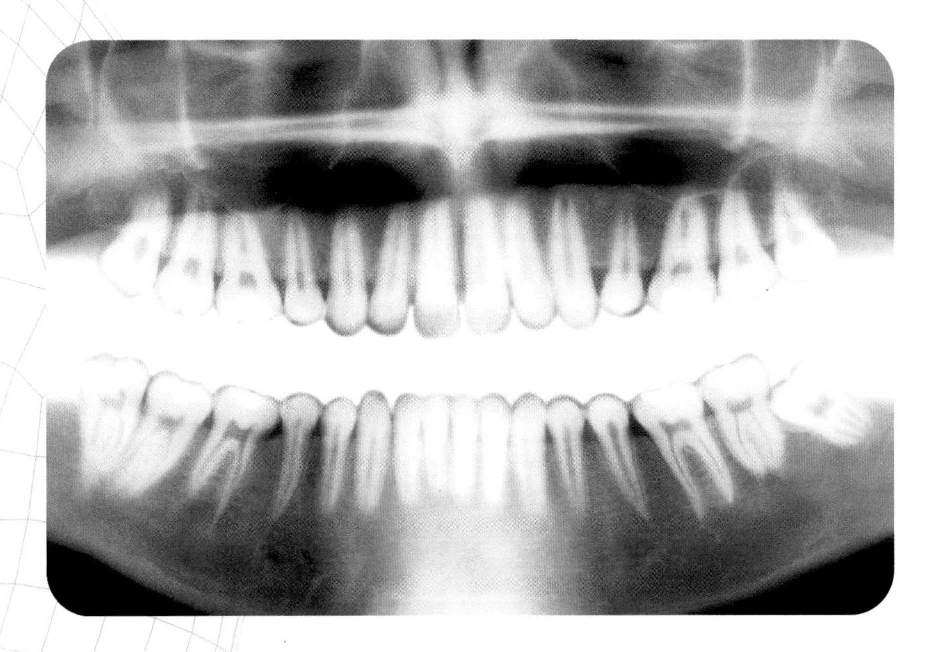

Say "Cheese!"

Your teeth are set into sockets in the jaw bone. Each tooth has an outer shell of enamel—the hardest substance your body makes. This tough layer surrounds another hard substance called dentine (similar to bone) and a soft central pulp cavity. Depending on its position in your mouth, each tooth has a particular function. Your front teeth (incisors and canines) cut and tear food into bite-sized chunks. Then, your back teeth (premolars and molars) grind it into a pulp, ready for swallowing.

The tingling sensation that you get from a carbonated drink is actually pain. The carbon dioxide gas, which makes the drink fizz, turns to acid in your mouth, stimulating the pain receptors on your tongue.

What is saliva?

The vast majority—99.4 percent—of your saliva is water; other components include enzymes, salts, and molecules that maintain a constant pH of 6.8. This slightly acidic environment provides the perfect conditions for the initial stages of digestion. Saliva is essential for tasting: It dissolves chemicals from the food or drink, which your taste buds perceive as flavors.

The taste sensation

The sensation of flavor, produced, in part by your taste buds, allows you to distinguish between foods. A typical adult has about 10,000 taste buds—the bulk of them on the tongue, with a few scattered around the back of the mouth. Each bud is specialized to detect one of four primary taste sensations: salty, sweet, sour, or bitter. Your sensitivity to tastes varies: For example, your taste buds are 1,000 times more sensitive to sour tastes than to sweet or salty ones and 100 times more sensitive to bitter tastes than to sour ones.

The throat and esophagus

The throat is a passageway for several different substances, including food and drink. These must be directed to the esophagus, which carries them to the stomach at a rate of up to 1½ inches per second.

A PASSAGE TO YOUR STOMACH

Your esophagus is a 10-inch-long muscular tube stretching from your throat to your stomach. Once you have chewed a mouthful of food, a precise sequence of swallowing maneuvers directs the ball of mush, or bolus, first to the throat and then into the esophagus. Food is moved along your esophagus by a muscular activity called peristalsis—a rhythmic sequence of contraction and relaxation of bands of muscle. The esophagus empties into the stomach at an opening called the cardiac orifice. On average, food reaches your stomach about 6 to 8 seconds after you swallow it, whereas liquids whiz through in 1 second flat. Rings of muscle called sphincters constrict the top and bottom of the esophagus to prevent stomach contents from going backward into the throat or mouth. These are called simply the upper and lower esophageal sphincters.

Traveling in reverse

Unfortunately, the esophagus does not always experience a one-way downward flow of traffic. The stomach contents are sometimes forcefully ejected by vomiting—an involuntary reflex that, although unpleasant, can serve a protective function. Vomiting helps to get rid of potentially dangerous substances very rapidly, ejecting poisons before they can get into the bloodstream.

The epiglottis—keeping food on the right track

Imagine that you are a piece of food that has just been swallowed and is en route to the stomach. Take a wrong turn in the throat, and you could end up going down the wrong tube. Two passages lead off the throat—the trachea and the esophagus. When you swallow a mouthful of food, the epiglottis closes over the end of the trachea until the food has passed by—a bit like railway switches directing a train down a particular track. If a morsel does go down the wrong way, you choke and cough sharply to bring it back up to the throat to be re-swallowed.

The epiglottis *is a flap of tissue that directs food and drink down the esophagus. Here it is in its normal position, allowing air into the trachea.*

The trachea, *or windpipe, channels air to the lungs.*

The esophagus *is a thin muscular tube, which is normally squashed flat by its position between the trachea and the spine.*

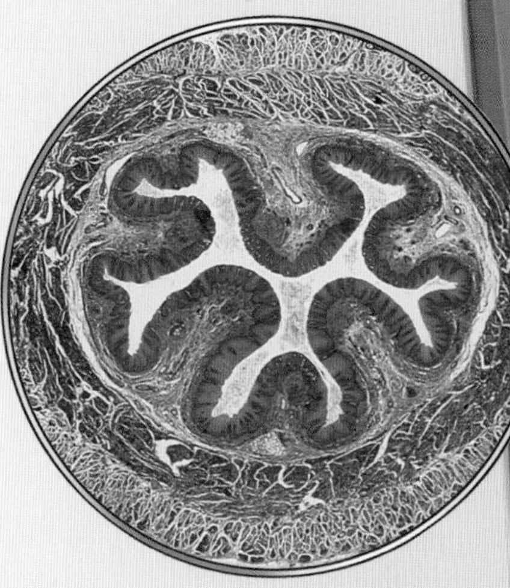

The esophagus in cross section
This micrograph shows the various layers in the esophagus wall. Its highly folded interior allows it to stretch around mouthfuls of food.

The cardiac orifice *is where the esophagus empties into the stomach.*

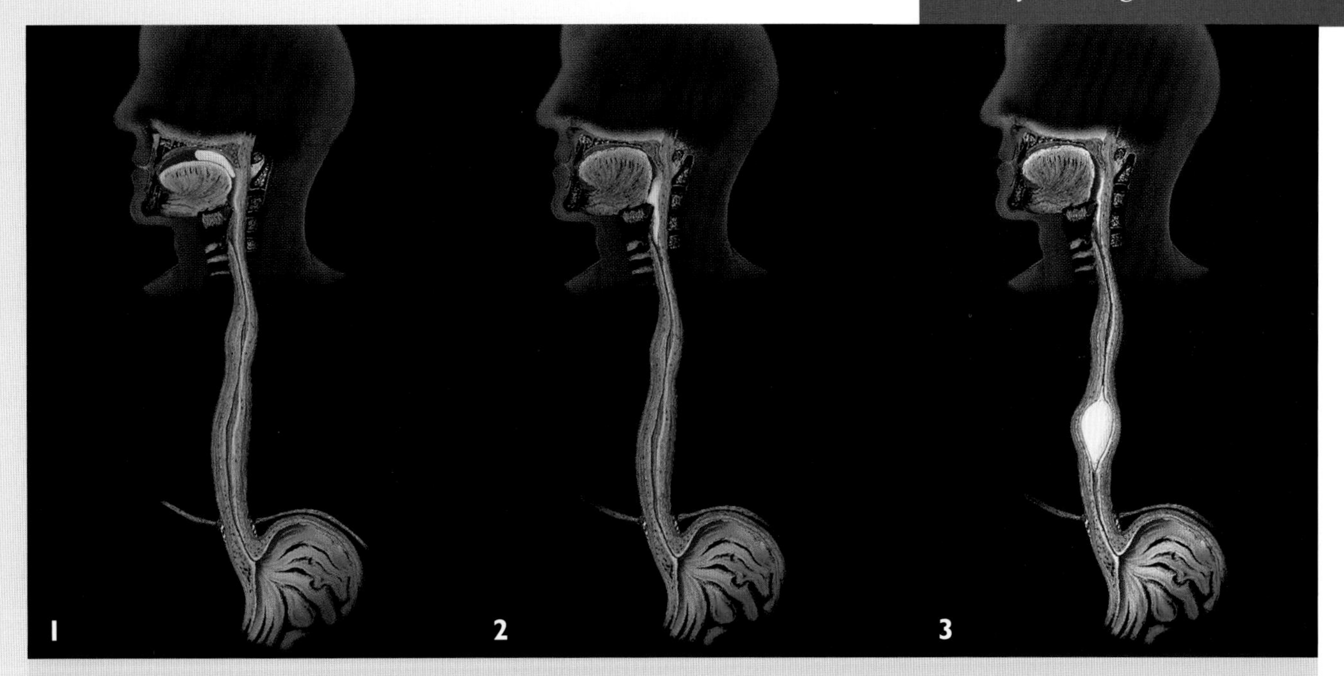

Down the hatch—the three phases of swallowing

1 Buccal phase The swallowing sequence starts voluntarily—in other words, it is under your control. When you are ready to swallow a bolus of food, your tongue pushes it up against the roof of your mouth and toward your throat. The combined mass of tongue and food raises your soft palate, shutting off the nasal space from your throat, so that food or drink doesn't go up the back of your nose. The bolus is then pushed into your throat, and the rest of the sequence proceeds involuntarily.

2 Pharyngeal phase As the bolus travels down your throat (the pharynx), your epiglottis folds over the opening of the trachea to prevent food from going down the wrong way. Between 0.2 and 0.3 seconds after you began to swallow, your upper esophageal sphincter relaxes to admit the bolus into the esophagus.

3 Esophageal phase Peristaltic waves of contraction and relaxation within the esophagus walls carry the bolus down toward the stomach. Once it arrives at the cardiac orifice, the lower sphincter opens to allow it to enter the stomach.

During swallowing, you actually stop breathing for a fraction of a second, as food travels past the closed entrance to the trachea.

The diaphragm *binds the top of the abdominal cavity; all other digestive organs are located below it.*

Peristalsis—muscular movement

Your digestive tract moves food along using muscles within its walls. Some muscles run lengthwise along the tract (longitudinal muscles), whereas others encircle it (circular muscles). In order to move food along, both muscle types work together in a mechanism called peristalsis. A wave of contraction passes along the tract, and this pushes a mass of food ahead of it. At the same time, a wave of relaxation moves along in front, lessening resistance to the food's passage. Peristalsis is not only a human mechanism; it is well illustrated by the way snakes swallow prey whole. This X-ray shows a frog en route from the snake's mouth along its digestive tract.

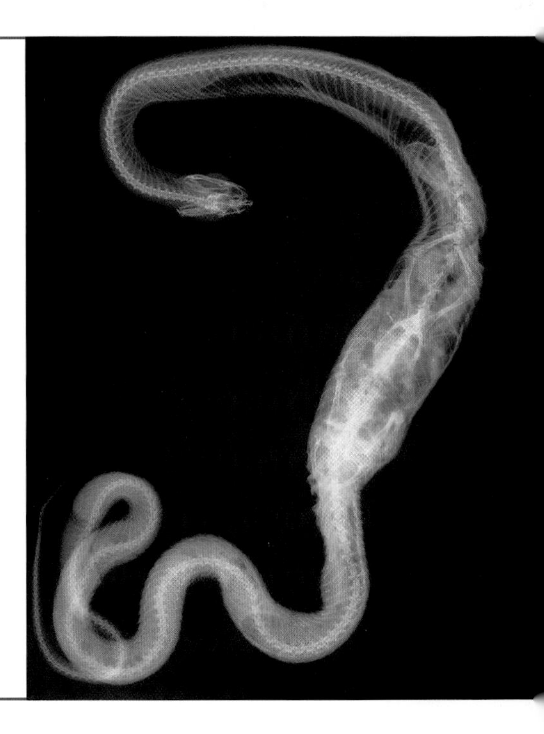

The stomach

The real business of digestion gets underway in the stomach, an inflatable bag that shrinks and grows with how much you eat and drink and that operates like a combination food processor and acid bath.

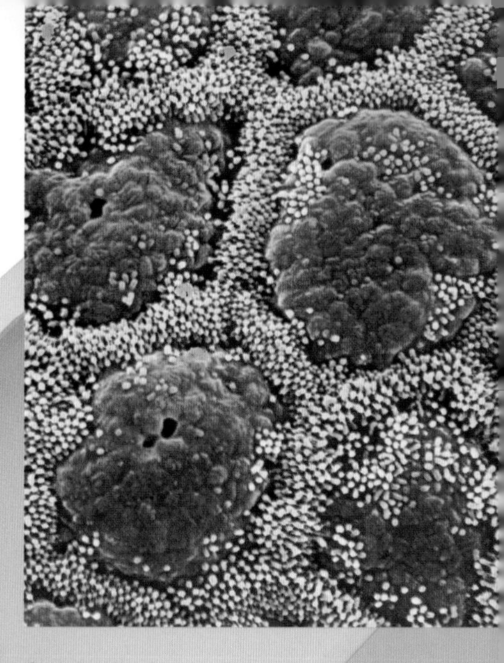

Gastric glands
This colored scanning electron micrograph shows a highly magnified view of the surface of the stomach lining. Here, globules of protective mucus (yellow) surround the mucus-secreting cells (red-brown).

TURNING YOUR STOMACH

Your stomach is a muscular J-shaped bag that has a smooth outer surface and a heavily folded lining, rich in gastric glands that produce digestive fluids and mucus. At the top of the stomach are the cardia, where food enters from the esophagus, and the fundus, which is heavily endowed with secreting glands. The outer curve of the "J" makes up the bulk of the stomach, known as the body, and at the bottom of the stomach is the antrum. The pylorus connects the stomach to the duodenum (the first part of the small intestine).

Flexing its muscles

Your stomach completes the mechanical breakdown of food, which was started by your teeth. Within the walls of the stomach are three bands of muscle, each oriented in a different direction, which enables your stomach wall to flex in three dimensions at once. This gives it the ability to churn, kneading and mixing food and gastric juices together into a thick, soupy substance called chyme. Your stomach also acts as a "storage tank" for food: An entire meal can be taken on board, churned into chyme, and then dispensed into the small intestine a little at a time. This slowing down of the chyme is important because the intestines work best on relatively small loads; if a meal were to flood through at the same rate at which you ate it, your intestine would not have time to process it properly.

The acid house

Glands in your stomach lining secrete hydrochloric acid, a powerful acid that helps break down all but the toughest components of food. It also provides a valuable line of defense against germs—few can survive this hostile environment for long. Many people are familiar with the sensation of heartburn—this is when stomach acid leaks back into the esophagus. To prevent the corrosive fluid from eating away at the stomach, the stomach lining is protected by a mucus layer, which contains a neutralizing agent (bicarbonate). If the mucus layer is damaged, however, say by bacterial invasion or by excessive alcohol, an ulcer may form.

The pyloric sphincter
opens and closes to control the flow of chyme, letting it through in small squirts.

The ring masters

Movement of food through the pyloric outlet is controlled by a band of muscle—a sphincter. Normally, this muscle is contracted, preventing material from passing through. When the sphincter relaxes, the outlet opens to allow some chyme to flow out of the stomach and into the duodenum. Other sphincters of the digestive system are located at the top and bottom of the esophagus; where the secretions of the liver, gallbladder, and pancreas feed into the duodenum; at the junction of the small and large intestines; at the end of the large intestine; and in the anus (two, internal and external). All except the last are involuntary: You have no conscious control over their opening and closing.

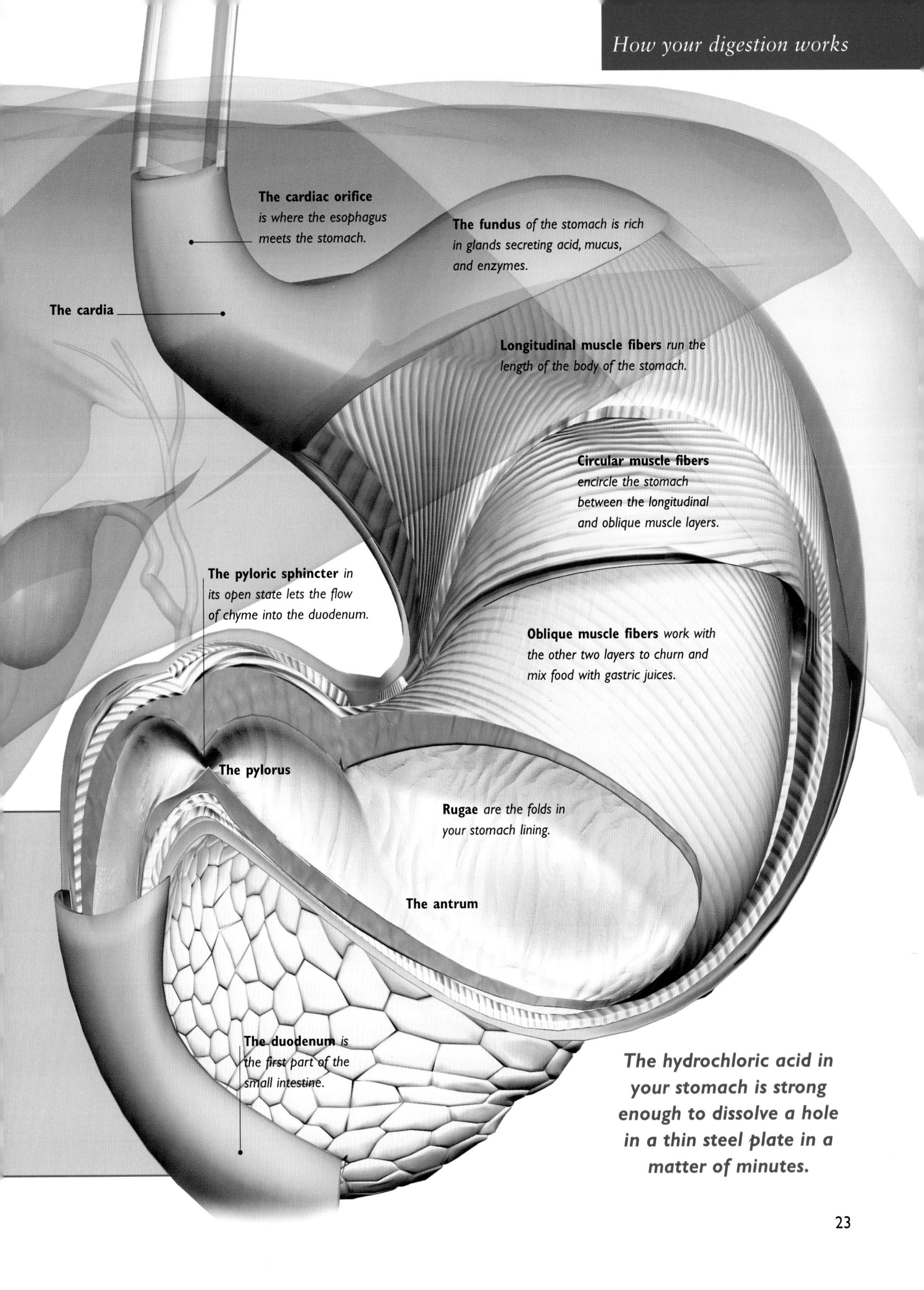

The hydrochloric acid in your stomach is strong enough to dissolve a hole in a thin steel plate in a matter of minutes.

The small intestine

Packed into your abdomen is the small intestine, the workhorse of the digestive system. This stretch of tubing, several yards long, is where the vast majority of digestion and nutrient absorption takes place.

The membrane that lines your abdominal cavity, the peritoneum, secretes 12 pints of lubricating fluid every day to prevent your intestines chafing against other organs as you move around.

THE LONG AND WINDING GUT

Eighty percent of nutrient absorption, and most of the chemical (rather than mechanical) digestion of food, occurs in your small intestine. The small intestine is a tube about 20 feet long and 1 to 1½ inches wide. In order to squeeze all of its length into your abdomen, the tube is folded repeatedly, winding back on itself several times. It is divided into three sections:

- **The duodenum** This is the first and shortest section of the small intestine, averaging about 10 inches in length. It acts a bit like a mixing bowl, blending the digestive juices from the liver, gallbladder, and pancreas with the soupy chyme from the stomach.
- **The jejunum** A sharp bend marks the beginning of this 7-foot, 6-inch-long section, where the bulk of the digestive and absorptive action happens.
- **The ileum** More digestion and absorption happen in this longest section—11 feet, 8 inches; it ends in a sphincter that leads to the large intestine.

The serosa *is the outermost layer of the intestine.*

RIDING THE WAVES

A cross section of the small intestine reveals a structure highly adapted for its role. Within the intestinal walls are both circular and longitudinal layers of muscle. Interleaved with these are networks of nerves, called plexi, which stimulate the muscles into three types of movement:

- **Segmentation** Rippling waves of contraction of the circular muscle churn the chyme on a stretch of intestine, ensuring that it mixes evenly with the digestive juices that are working on it.
- **Local peristalsis** Weak waves of peristalsis, limited to about an inch at a time, keep chyme moving slowly along the intestine.
- **Coordinated movements** Occasionally, larger-scale peristalsis shifts chyme in bulk, moving it several feet at a time, encouraging the release of more chyme from the small into the large intestine.

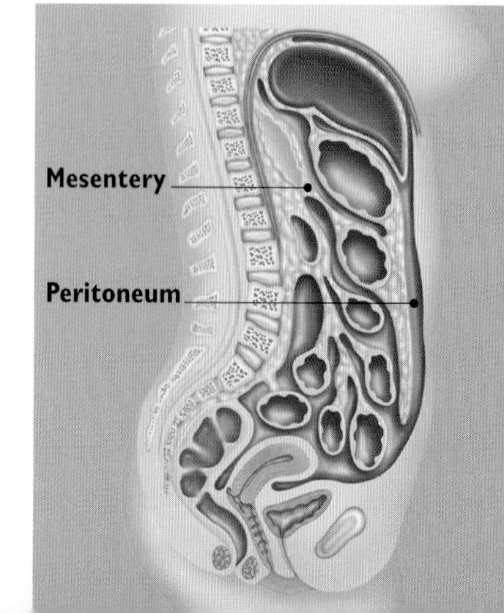

What's holding your GI system in place?

Your digestive tract and its associated organs are supported and held in place in your abdomen by the muscles of your back, abdominal wall, and diaphragm, and also by a set of membranes called mesenteries. These are sheets of connective tissue that anchor the tubes, bags, and organs of the digestive system to one another and also to the peritoneum—the membrane that lines the inside of the abdominal wall. Mesenteries also carry vital blood vessels and nerve fibers.

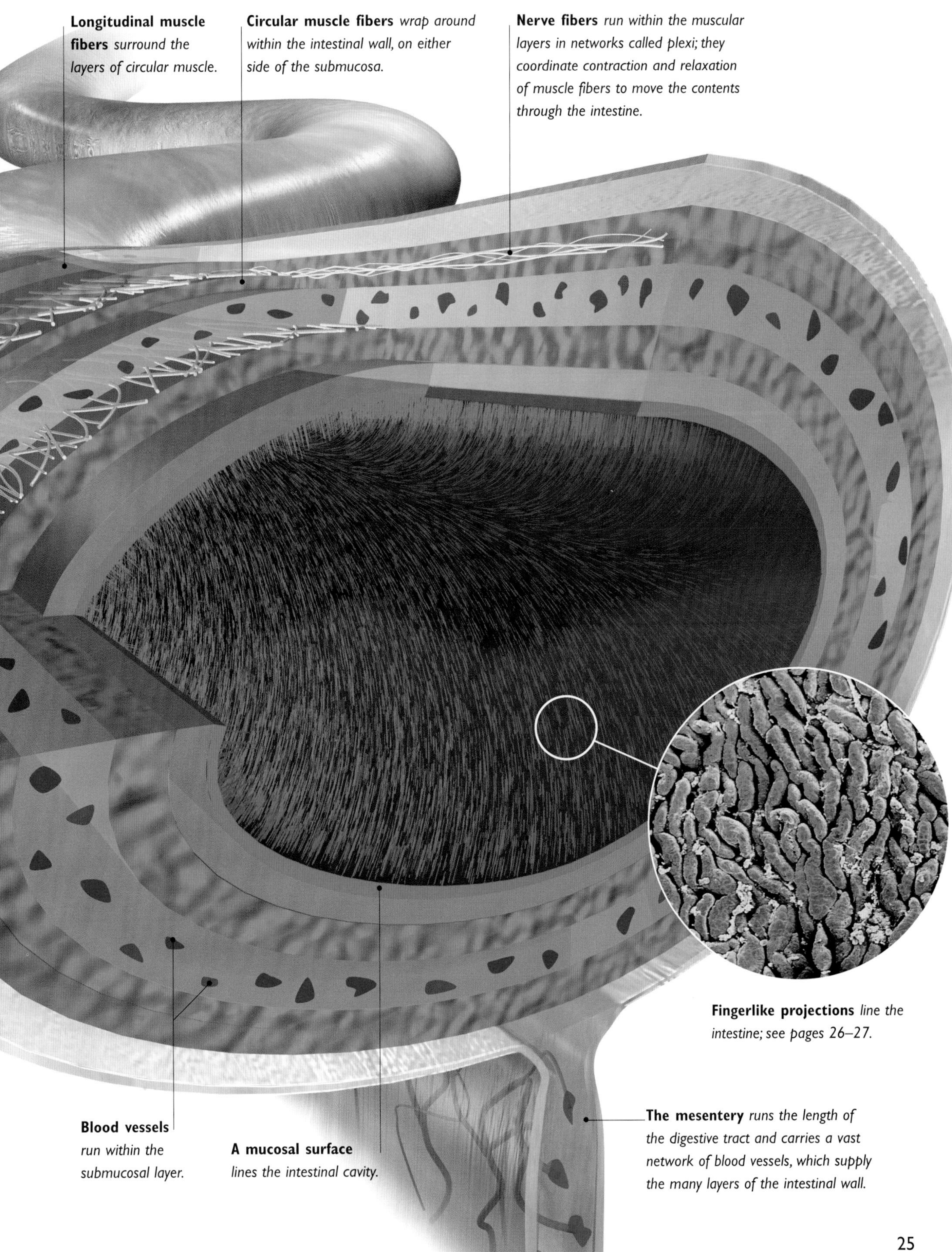
Longitudinal muscle fibers *surround the layers of circular muscle.*
Circular muscle fibers *wrap around within the intestinal wall, on either side of the submucosa.*
Nerve fibers *run within the muscular layers in networks called plexi; they coordinate contraction and relaxation of muscle fibers to move the contents through the intestine.*
Fingerlike projections *line the intestine; see pages 26–27.*
Blood vessels *run within the submucosal layer.*
A mucosal surface *lines the intestinal cavity.*
The mesentery *runs the length of the digestive tract and carries a vast network of blood vessels, which supply the many layers of the intestinal wall.*

The intestinal lining

To maximize the absorption of nutrients from the food and liquids that enter your GI tract every day, the lining of your intestine is a highly specialized environment of fluid-spewing crypts and towering nutrient-hungry columns.

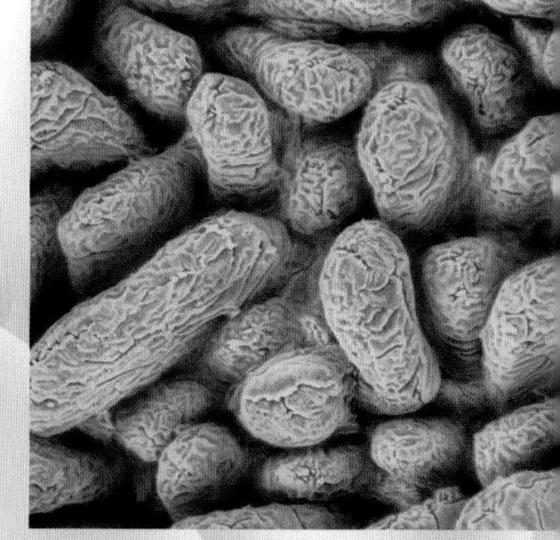

Wall-to-wall carpet
The velvety lining of the small intestine has many folds, each covered in millions of tiny, fingerlike projections called villi (seen here from above, magnified 85 times). The surface of each cell on the villi is studded with up to 1,000 tiny hairs called microvilli. In a ½-square-inch area of lining, there would be about 10,000 villi and 5,000 million microvilli.

AN ABSORBING TALE

Imagine that you are a particle of food, drifting through the small intestine. On all sides gigantic columns, the villi, rear above you, their surfaces coated with a rippling carpet of hairs. At the feet of the villi are strange holes, the intestinal crypts, which produce 3⅛ pints of fluid in the intestine every day. The surface of both the villi and the crypts are composed of two types of cell, enterocytes and goblet cells. Wherever they are, the goblet cells have one purpose: to secrete a layer of protective mucus that also lubricates the passage of the chyme. The enterocytes, however, have two roles. In the crypts, they secrete a watery fluid that helps to dissolve the nutritious products of enzyme digestion in the chyme; the enterocytes on the villi actively absorb the nutrient-rich fluid. Enterocytes are highly active cells that "burn out" and die in just five days.

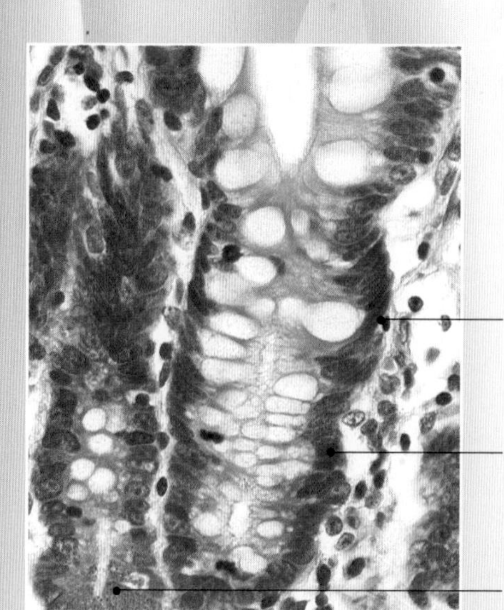

Down in the crypts
This micrograph shows a longitudinal section through an intestinal crypt, also known as a crypt of Lieberkühn. Toward the top, the crypt's entrance into the intestinal lumen is visible, bounded by the bases of two villi.

An enterocyte *manufactures and secretes the watery fluid vital to the absorptive functions of the villi.*

A goblet cell *makes protective mucus.*

A Paneth cell *may secrete an enzyme called lysozyme, but researchers are still not clear on all of its functions.*

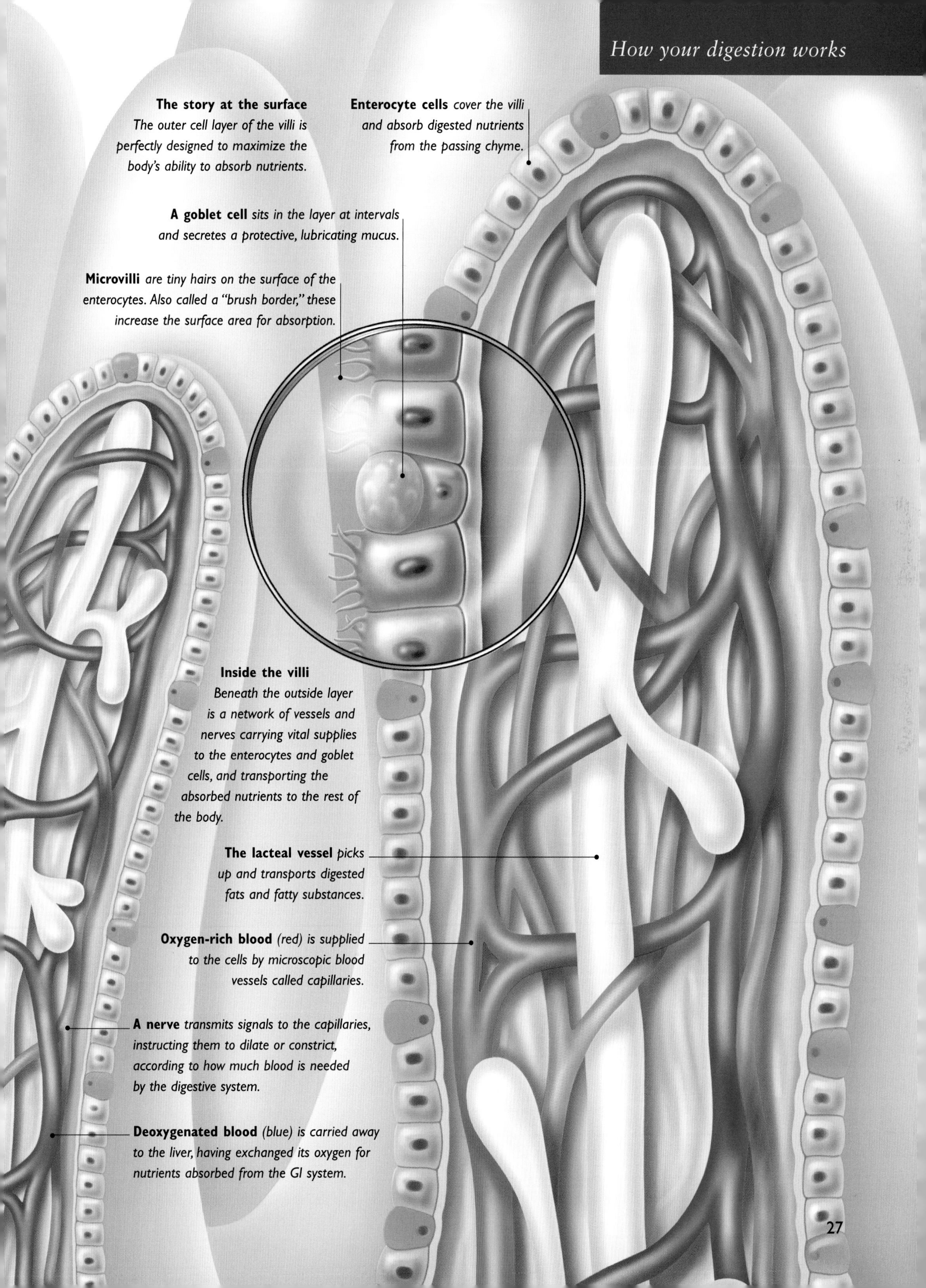

The story at the surface
The outer cell layer of the villi is perfectly designed to maximize the body's ability to absorb nutrients.

Enterocyte cells *cover the villi and absorb digested nutrients from the passing chyme.*

A goblet cell *sits in the layer at intervals and secretes a protective, lubricating mucus.*

Microvilli *are tiny hairs on the surface of the enterocytes. Also called a "brush border," these increase the surface area for absorption.*

Inside the villi
Beneath the outside layer is a network of vessels and nerves carrying vital supplies to the enterocytes and goblet cells, and transporting the absorbed nutrients to the rest of the body.

The lacteal vessel *picks up and transports digested fats and fatty substances.*

Oxygen-rich blood *(red) is supplied to the cells by microscopic blood vessels called capillaries.*

A nerve *transmits signals to the capillaries, instructing them to dilate or constrict, according to how much blood is needed by the digestive system.*

Deoxygenated blood *(blue) is carried away to the liver, having exchanged its oxygen for nutrients absorbed from the GI system.*

The liver, gallbladder, and pancreas

Although not part of the digestive tract that food passes through, the accessory organs—the liver, gallbladder, and pancreas—are essential to digest the large and complex molecules that make up food.

You can lose up to 80 percent of your liver and it will still function normally, regenerating the lost cells within about 10 days.

THE LIVER

Digestion is only one of the liver's areas of responsibility—the liver is the most versatile organ in the body, with more than 200 different functions. In digestion it controls the manufacture of bile, the collection and processing of nutrients, and vital metabolic functions.

- Manufacture of bile Bile is a green liquid made in the liver and stored in the gallbladder. It is mostly water but also contains buffers and bile salts. Buffers help to counteract stomach acid to keep the conditions in the small intestine at the right pH for enzymes to work on the chyme. Bile salts come into play on foods that are too much for enzymes alone, helping to break up large droplets of fat in the chyme into tiny ones. This process, known as emulsification, allows the digestive enzymes to break down the fats into molecules small enough to be absorbed in the small intestine.
- Collection and processing of nutrients The nutrients absorbed by your stomach and intestines have to be treated by the liver before they can be used by the rest of your body. A large blood vessel called the portal vein collects all nutrient-rich blood returning from the intestinal lining and takes it directly to the liver. Once processed, nutrients are fed back into the bloodstream to be circulated to the rest of the body.
- Metabolic functions The liver performs an array of life-supporting functions via complex metabolic pathways; these include manufacturing enzymes, breaking down toxins, removing alcohol from the blood, and storing energy in the form of a quickly accessible carbohydrate reserve called glycogen.

THE PANCREAS

Lying behind the stomach is the 6-inch-long pancreas. This produces a fluid called pancreatic juice that contains a range of enzymes—such as trypsin, which helps digest proteins—and buffers, which help neutralize stomach acid. Inside the pancreas, branches of the pancreatic duct collect the juice from hundreds of lobules—rather like a bunch of grapes—then join up into a single vessel that leads to the duodenum. In addition to its digestive functions, the pancreas manufactures hormones, such as insulin.

THE GALLBLADDER

Bile passes from the liver, via the cystic duct, to a small pear-shaped sac called the gallbladder. Up to 2½ fluid ounces of bile is stored here, and while it waits to be squirted into the duodenum via the bile duct, its composition changes. Water is absorbed by the walls of the gallbladder to concentrate the solution of salts—it can be up to 20 times stronger than the original bile manufactured by the liver. When a new load of chyme arrives in the duodenum, the gallbladder contracts and ejects its contents to help emulsify the fat globules.

The liver *is the largest and heaviest organ in the body at 3⅓ pounds.*

The gallbladder *stores and concentrates bile.*

The cystic duct *is the tube through which bile travels on its way to the bile duct and then the duodenum.*

The pancreas would digest itself in a few hours if it weren't for a special substance called trypsin inhibitor, produced to inhibit its own enzymes.

The pancreatic duct *transports the pancreatic juices to the duodenum.*

The pancreas *comprises lobules made from units called acini. These manufacture and secrete highly digestive fluids—about 2⅔ pints (10 times its own weight)—into the pancreatic duct every day.*

The bile duct

The duodenum

The ampulla of Vater *is the "meeting point" in the duodenum where the bile and pancreatic ducts empty their contents.*

The sphincter of Oddi *controls the release of the fluids in the now merged bile and pancreatic ducts.*

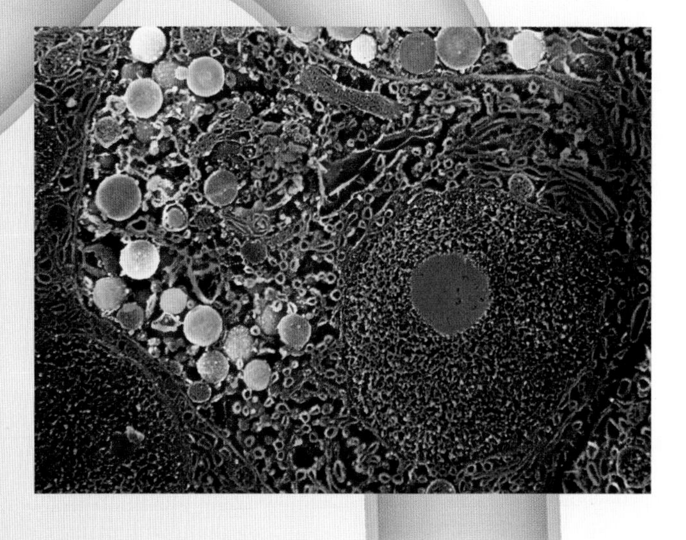

Up close in the pancreas
This scanning electron micrograph of a pancreatic acinar cell shows zymogen granules (yellow, brown, and white circles), which are the inactive form of a digestive enzyme. The purple folds are membranes that synthesize enzyme proteins.

The biochemistry of digestion

Before your body can absorb the nutrients of your last meal, the bonds that hold their molecules together must be broken. Your body employs an army of special proteins called enzymes, which work at different stages of the digestive process.

BIOCHEMISTRY BASICS

Food is made up of three basic nutrients—proteins, carbohydrates, and fats. Each is a large molecular structure made from many repeating smaller units.

- **Proteins** Every protein molecule is made from amino acids. A molecule of two amino acids is called a dipeptide, whereas a long chain of amino acids is a polypeptide. A complex molecule containing many polypeptides is a protein.
- **Carbohydrates** The two basic groups—sugars (simple) and starch (complex carbohydrates)—are made from the same basic unit: a monosaccharide. Sugars contain one or only a few monosaccharides joined together; glucose, for example, is a monosaccharide, and sucrose is a disaccharide. Starch molecules are complex because they comprise long chains of polysaccharides.
- **Fats** These are made up of two basic units—glycerol and fatty acids.

Enzymatic digestion

In order to break the long chains of nutrient molecules into units small enough for your intestines to absorb, you need some means of "snipping" the bonds holding the units together. This is where your digestive enzymes come in. These complex proteins latch onto a target substance, break it into smaller pieces, and then move on to the next target. Each enzyme works on one substance only; for example, the enzyme dipeptidase chops up dipeptides into amino acids. Furthermore, different enzymes work at different locations: Amylase works on starch in the mouth, pepsin works on proteins in the stomach, and pancreatic lipase breaks up fats in the small intestine.

Sugar crystals
Probably the most familiar form of sugar in everyday life, sucrose (shown here in crystal form, magnified 60 times) is a disaccharide, that is, two monosaccharide molecules joined together.

What happens where?
Join us on a digestive journey to see how large molecules are gradually transformed into smaller ones, using the visual key below.

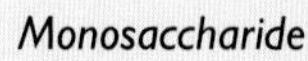

Monosaccharide
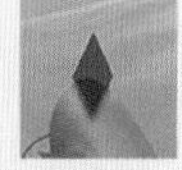

Starch

Amino acid
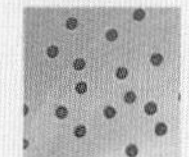

Protein

Fatty acids and glycerol

Fat droplet

Amylase

Stomach acid
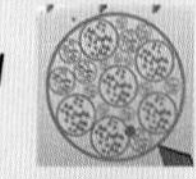

Protein-attacking enzyme

Bile salt

Lipase

Water

Fiber

A bit of a mouthful
Enzymatic digestion begins in the mouth. Your saliva contains an enzyme called amylase, which starts the breakdown of starch.

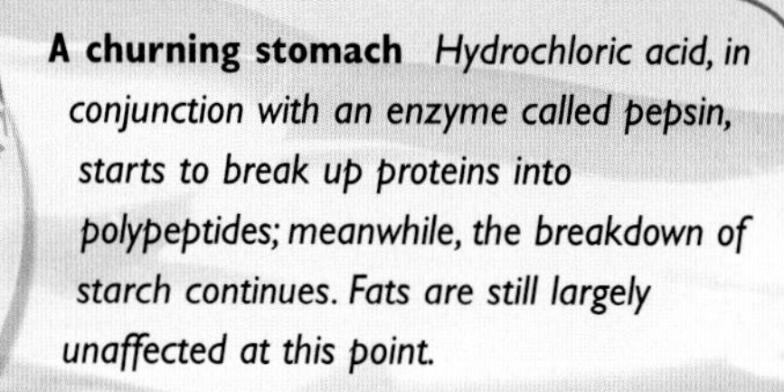

A churning stomach *Hydrochloric acid, in conjunction with an enzyme called pepsin, starts to break up proteins into polypeptides; meanwhile, the breakdown of starch continues. Fats are still largely unaffected at this point.*

Starting off in the small intestine
The main action is in the duodenum. Starch is broken into short chains and glucose units, and protein-attacking enzymes convert polypeptides into dipeptides and amino acids. Bile salts emulsify the fat into tiny droplets, small enough for the enzyme lipase to work on.

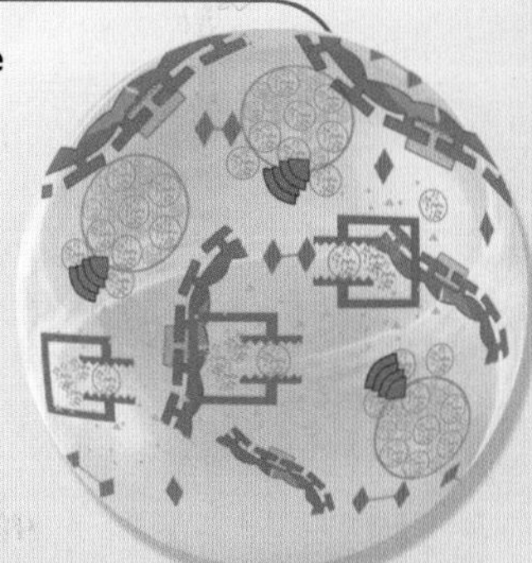

Finishing the job *The breakdown of starch into glucose and other monosaccharides is completed by the enzymes lactase, sucrase, and maltase. Trypsin, chymotrypsin, and dipeptidase, among others, turn poly-peptides into amino acids. Lipases have broken up fat droplets into units of fatty acid and glycerol, which are repackaged in water-soluble form. The basic units are now ready to be absorbed by the intestinal villi.*

The end of the tunnel
When the chyme is ready to pass into the large intestine, very little is left except water, some vitamins and minerals, and insoluble fiber.

The large intestine

Home to a thriving population of "friendly" bacteria and the occasional bubble of pungent gas, your large intestine extracts the last drops of goodness from food and prepares whatever is left for departure.

Friend or foe?
Escherichia coli *bacteria are best known for causing outbreaks of food poisoning. But not all strains of* E. coli *are bad for your digestion; some strains, in fact, form a major component of a healthy gut's flora.*

GROSS ANATOMY

The final part of your digestive tract, the large intestine describes a 5-foot circuit of your lower abdomen, framing the small intestine. Its main functions are the absorption of water, vitamins, and minerals, and compacting into feces what is left from the digestion process. It is divided into three main parts—the cecum, the colon, and the rectum (see below).

Look who's coming to dinner

Thousands of millions of bacteria, collectively known as your digestive flora, live in harmony in your large intestine. The bacteria feed off the material that passes through your GI system, and in return they perform a number of important roles: They produce vitamins, including B_{12}, folic acid, and thiamine, some of which your body cannot make for itself; they help to break down bile pigments, so they can be reabsorbed and returned to the liver, where they are reassembled and reused (leftover pigments give stool its color); through competition for food with other bacteria, digestive flora play a vital defensive role; and they feed off fiber that is too tough for digestive enzymes, producing fatty acids small enough for you to absorb—an important source of energy.

Gas power

Your digestive tract usually contains about 5 fluid ounces of gas; some of this is swallowed air and some results from the breakdown and fermentation of food. Most gas is absorbed through the intestinal walls into your bloodstream, but occasionally some builds up, causing discomfort. Gas in the lower intestine is mainly composed of hydrogen, carbon dioxide, and methane, all of which are odorless. Unpleasant smells come from byproducts of bacterial protein metabolism, such as hydrogen sulfide, indole, or skatole.

There are 10 times more bacteria in your digestive tract than there are cells in the rest of your body.

A guide to the large intestine

a The cecum forms a pouch at the start of the large intestine. Here chyme is collected, fed in from the small intestine via the ileocecal valve, and the process of compaction begins. Leading off the blind end of the cecum is the appendix, a slender 3½-inch tube dotted with lymph nodules, where white blood cells congregate to fight germs.

b The colon is the main body of the large intestine. It is subdivided into four regions—**(1)** the ascending, **(2)** transverse, **(3)** descending, and **(4)** sigmoid colon. Three ribbons of muscle that run along the colon—the tenia coli—pull the walls into a series of pouches called haustra, allowing it to expand when a large load is passing through.

c The rectum is the last 6 inches of the digestive tract; it is a muscular tube that stores feces before it is excreted. The urge to defecate is triggered when material arrives from the sigmoid colon, but the exit from the rectum is controlled by two sphincters—the internal and external anal sphincters. The latter is under voluntary control.

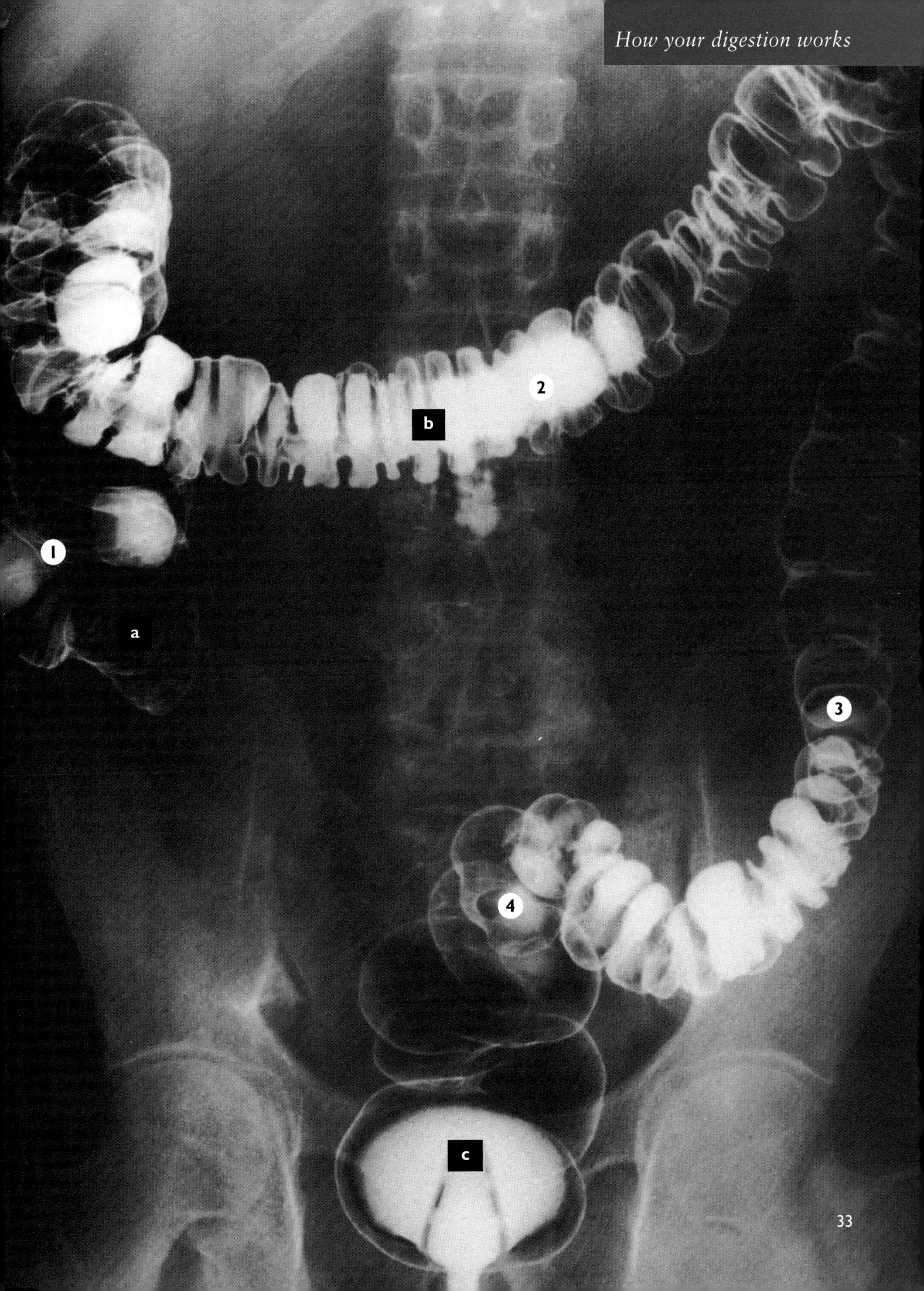
1
2
3
4
a
b
c

A day in the life of the digestive system

Only the top and bottom of the digestive tract are under your conscious control. The rest depends on complex interactions between your unconscious brain and the different parts of the digestive system.

THE DAILY GRIND

The length of time it takes for a meal to work through your GI system depends on a host of factors—the content and timing of the meal, the state of your digestive tract, how active you are, and how much you've eaten recently. On average, though, food comes out of one end between 14 and 24 hours after it went in the other. For your digestive system to extract all the nutrients from a meal, its grinding, squeezing, mixing, secreting, and compacting are carefully coordinated by a system of nervous and hormonal signals. Here, we follow the journey of your breakfast throughout the day.

7:30 A.M. Hungry? Feed me now

It's been 12 hours since you last ate, and receptors monitoring your blood sugar levels detect that you need rapid refueling. Part of your brain, called the hypothalamus, gathers this information and sends signals to the conscious portions of your brain, producing the sensation of hunger. Your appetite drives you into the kitchen. The sight and smell of your partner's fresh toast activate some of the primitive parts of your brain, producing a barrage of nerve messages that literally make your mouth water and trigger the secretion of gastric juices in your stomach.

12:00 A.M. Chyme at midnight

Most of what is left of your breakfast has now reached your sigmoid colon. Arriving home late, you treat yourself to a midnight feast. As food enters your stomach, the gastrocolic reflex triggers a wave of peristalsis that pushes the remains of a ball of chyme—by now a stool—into your rectum. Stretch receptors in the rectal wall send messages to your brain, producing the urge to defecate. They also signal the internal anal sphincter to relax, allowing the stool to move into the anorectal canal. You visit the bathroom to relax your external anal sphincter, and then go off to bed.

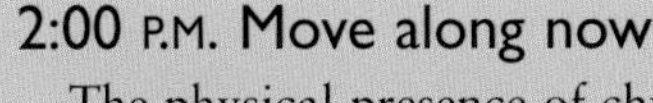

2:00 P.M. Move along now

The physical presence of chyme in the small intestine excites and irritates its walls. In response, the intestinal muscles twitch and contract, producing local movements that slowly mix and shift chyme along. By now the first masses of chyme from breakfast arrive at the cecum. Material moves very slowly from here, via powerful peristaltic waves triggered by your body clock and the arrival of a new meal in your stomach.

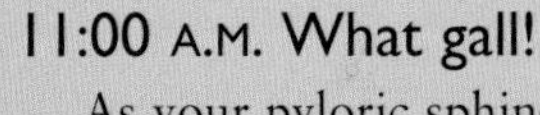

11:00 A.M. What gall!

As your pyloric sphincter squirts chyme into your duodenum, the acidic gastric juices stimulate hormonal signals that trigger the pancreas to produce its alkaline juices to counteract the stomach acid. When the fattier parts of your breakfast arrive, they trigger the production of a different hormone, which prompts the gallbladder to release bile.

7:45 A.M. A hearty breakfast

At last you sit down to breakfast: muesli with milk, croissant, and coffee—a meal that provides carbohydrates in abundance, as well as some fat and protein. You eat, and receptors in your stomach wall detect the size and content of the meal. Signals to your brain produce an almost immediate loss of hunger (although not necessarily of appetite), while your stomach itself responds by flexing its muscles to churn your meal into chyme and by secreting acid and enzymes.

2 Healthy digestion for life

LOOKING AFTER YOUR DIGESTIVE SYSTEM

In 400 B.C. Hippocrates wrote that "a bad digestion is the root of all evil." For most of us, however, the gastrointestinal (GI) system is simply something that we take for granted—until it malfunctions. More than half of us will suffer from digestive problems at some stage in our lives, yet many of these ailments can be avoided. By developing greater awareness of the health of your GI system, what is good for it and what isn't, you can help keep your digestive tract trouble-free and minimize any problems that might occur.

Be aware of what is normal for you and your family, and be alert to any changes that may indicate a problem.

Learn how to control your risk factors by eating well, not smoking, and maintaining good food hygiene.

Work with your physician to decide if you need to make serious lifestyle changes to prevent or treat digestive illness.

Be aware of what is normal

Learning about your body's day-to-day workings is a great way to become more self-aware. By knowing what is normal for you, you can be alert to any unusual changes that might indicate a problem.

As with many aspects of looking after your health, when it comes to GI system health, awareness is the key: knowing what is normal and what is not. After getting food into your mouth, digestion is an unconscious process, so you are not really aware of it. The best bet, then, is to be on the lookout for unusual symptoms. An abdominal symptom could indicate that something is wrong (for more information, see pages 50 to 53), but bowel habit is probably the most obvious indicator of how well the digestive system is working.

WHAT'S YOUR HABIT?

We're all unique when it comes to bowel habits. "Regularity" may mean bowel movements twice a day for some people or just twice a week for others—there is no "right" number.

Look out for changes

Although variety is normal in the bowel habits of the population in general, when it comes to your own bowel movements, a change may indicate a problem. Try to take note of the usual frequency and pattern of your own trips to the toilet, and be aware of the size, amount, and consistency of the stools.

Some people are always "regular," for instance they always pass a stool soon after getting up in the morning or after eating. But for others, bowel habits vary considerably. Diet has the most obvious effect—a particularly spicy meal, for example, produces a familiar change in bowel habit. In general, the more fiber you eat, the larger and more frequent the stools.

It's not just the food you eat that influences your bowel habit; the amount you drink—especially water, coffee, and alcohol—also has an impact. So, too, do cigarette smoking and physical activity. Even your mood influences your trips to the bathroom: Anxiety, for instance, tends to lead to "the runs," whereas depression may cause constipation.

Women often find that their bowel habits change when they are menstruating, during pregnancy, or at menopause. This is caused by changing levels of the sex hormones estrogen and progesterone, both of which affect bowel function. Prescribed and over-the-counter medicines can also cause diarrhea or constipation (see also page 50).

A CHANGE FROM THE NORM

How can you tell if you really are constipated or have diarrhea? These terms often mean different things to different people. When doctors talk about constipation, they mean infrequent motions—fewer than two

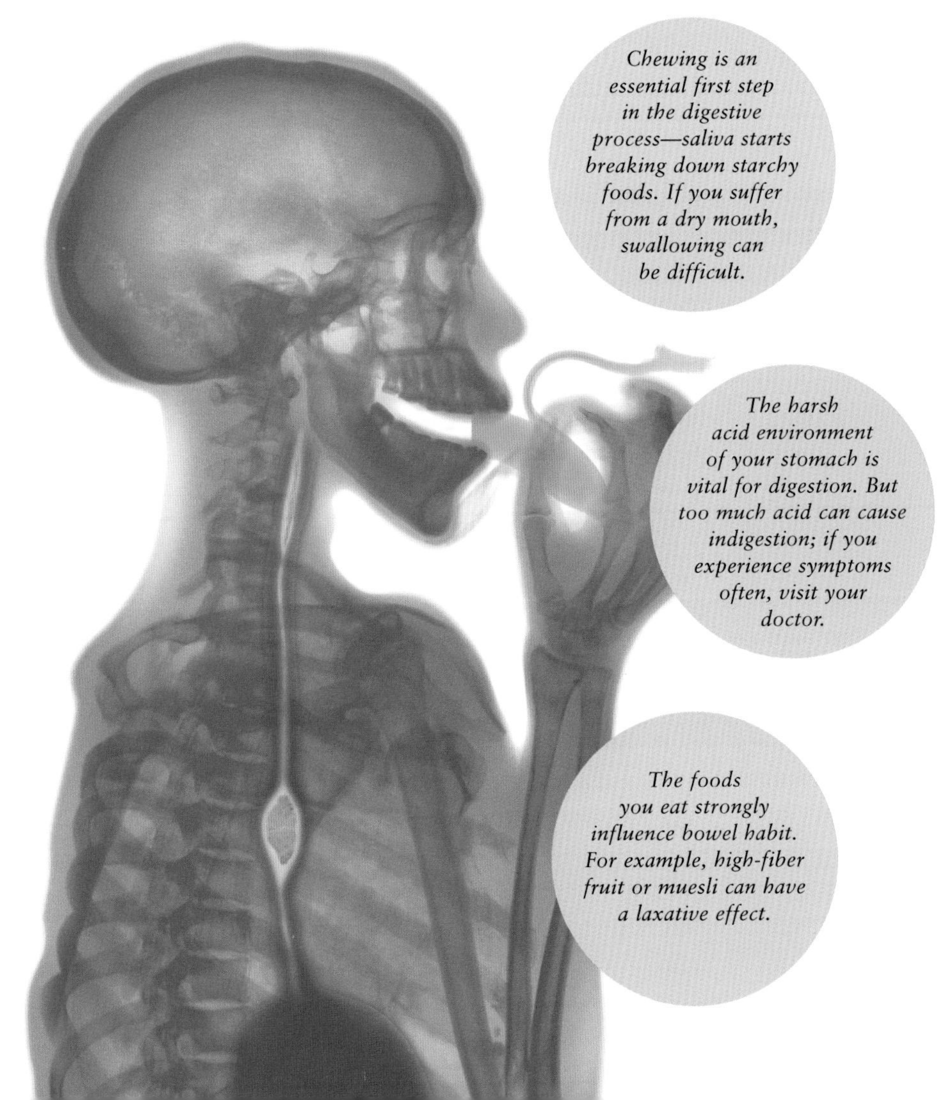

Digestion—an unconscious process *Once we have swallowed food and it enters the stomach, digestive enzymes and acid break it down into components that can be absorbed and used by the body.*

Healthy teeth, healthy GI system

Dental hygiene is important for your GI system's health—the entire process of digestion begins with chewing and the action of saliva on food. If you look after your teeth and gums, you probably won't need to worry. But if you think your dental hygiene routine is inadequate, ask your dentist or dental hygienist for advice on brushing technique and antibacterial products.

per week—or stools that are small, hard, or difficult to pass. Diarrhea means bowel movements that are very loose or frequent—in other words, more than three times per day. A change in bowel habit, especially if it lasts for more than three weeks, is more worrisome than the absolute frequency of bowel movements.

COMMON DIGESTIVE PROBLEMS

Among the most common conditions affecting the digestive system are constipation, gastroenteritis, and irritable bowel syndrome.

Constipation

Women tend to be troubled by constipation more than men, and elderly people are particularly susceptible because of weakening muscles in the colon and pelvis as they age. Usually, the cause of constipation can be pinned down to a low-fiber diet, physical inactivity, long-term laxative use, or a combination of the three. In people under 50, constipation is rarely a sign of anything serious. After this age, however, the likelihood of a more serious cause, such as a blockage caused by a tumor, is greater. If you are constipated a lot, it's worth seeing your doctor about it.

Constipation is a common cause of piles, also known as hemorrhoids. These usually improve if you increase your dietary fiber and fluid intake. But if you have persistent bleeding from the rectum, it's important to see your doctor so that a more serious condition, such as colon cancer, can be ruled out.

Acute gastroenteritis

This infectious condition causes abdominal pain, diarrhea, nausea, and vomiting. It usually begins suddenly, lasts for one to several days, and then clears up on its own. The usual culprits behind a bout of gastroenteritis are bacteria (sometimes from food) and occasionally viruses (see page 43). Bloody diarrhea and dehydration are worrisome symptoms and should prompt a doctor's visit.

Irritable bowel syndrome

Other than constipation, irritable bowel syndrome (IBS) is the most common complaint affecting the intestines. About 7 percent of men and 15 percent of women have the condition at some time; symptoms include abdominal pain, bloating, and altered bowel habit. The condition may be triggered by a GI infection or antibiotics, and it is often worse when under stress (see pages 52–53). IBS may last for years and can disrupt social life, travel, and mood. Fortunately, though, IBS does not bring with it a higher risk of cancer.

UNDER-12s

A tummyache or something more serious?

Abdominal pain is very common in children. It is not usually a sign of anything serious, but it could have a number of different causes.

If your child has a tummyache, it might help to ask yourself the following questions:

- Has the pain come on before or after a stressful situation connected with school, friends, or parents?
- Is the pain similar to any previous episodes?

In these cases it's unlikely that there is a serious underlying cause.

The presence of any of the following symptoms, or a new type of pain, however, should prompt you to call the doctor:

- high fever or a racing pulse;
- being much quieter than usual or suddenly listless;
- a tender-to-the-touch abdomen;
- other associated symptoms, such as pain going to the bathroom, vomiting, diarrhea, constipation, or abdominal swelling.

Is chewing gum good for your teeth?

Chewing gum boosts the amount of saliva produced after eating—a good thing for two reasons: (1) Saliva contains bicarbonate to neutralize the acid produced by dental plaque, and (2) it is antibacterial. So it's worth chewing (sugar-free) gum after a meal for the health of your teeth. Chew on both sides of the mouth to release saliva on both sides. The downside to chewing gum is that it encourages your stomach to make acid—you fool it into expecting food—and you may feel bloated. If you have stomach ulcers, it's best to avoid it.

ASK THE EXPERT

Diverticular disease

This disorder, in which small pockets protruding from the lining of the large intestine become inflamed and infected, can cause abdominal pain, diarrhea, or even passage of blood. About one in every three people over the age of 60 has diverticular disease. Unlike IBS, though, it's more common in middle age and beyond. Diverticular disease mainly affects people living in developed countries, and experts believe it results from relatively low-fiber diets combined with the lowered muscle strength that accompanies aging.

Kiddie tummy trouble

About 10 percent of children have recurrent stomach pain—and in most cases no physical cause is found. The most common causes are stress or constipation, but there may be a more serious cause, such as gastroenteritis (with diarrhea or vomiting), urinary tract infection, IBS, and appendicitis.

HAPPY TOILET TRAINING

Two tight bands of muscle around the anus, called anal sphincters, are responsible for preventing fecal incontinence. The internal sphincter relaxes on its own when the rectum fills up with feces. The external sphincter is under your control and it is this that you keep squeezed tight until you find a toilet. Babies and young children have no control and respond to the sensation of a full rectum by immediately soiling their diapers. It's not until the second year of life that toddlers gradually learn to exercise control over their external sphincter, usually before they control the bladder. The next stage is potty training, where the child must develop the social skill of using a toilet. Don't be tempted to start training before your child is ready, and be prepared for accidents; try to maintain an attitude of relaxed encouragement. A delay in gaining control of the bowels is common; there should be little concern until about four years of age. If your child still has no bowel control after this age, see a doctor.

TRIPS TO THE TOILET

The cardinal rule regarding toilet habits is to follow the dictates of your body, not dictate to it. The intestines are a sophisticated and complex but finely tuned system that works like clockwork. It will tell you when it needs emptying. Don't feel that you must take a trip at a certain time each day. On the other hand, the urge to have a bowel movement should be obeyed within a short time wherever possible—constantly delaying such signals can result in constipation. It is also important to avoid straining to excess. If you do, you'll be increasing the pressure in the rectum and you're more likely to develop hemorrhoids.

On the potty
Potty training can start as soon as a child has developed conscious control of his or her bowels. This normally happens some time during the second year of life—before this time, attempts at potty training are futile.

PROTECT YOUR GI SYSTEM

FILL UP ON FRUIT AND VEGETABLES Eating plenty of fresh produce can prevent and counteract all sorts of GI problems, from constipation to cancer.

TACKLE STRESS AND LEARN TO RELAX Some digestive problems are linked to stress, so try yoga or massage to reduce the pressure.

EXERCISE REGULARLY A sedentary lifestyle is bad news for your digestive health. Build activity into your everyday life and you'll soon reap the digestive benefits.

Control your risk factors

What steps can you take to safeguard the health of your digestive system? The first thing you can do is to look at whether you need to make lifestyle changes. Then it's just a matter of taking sensible precautions when dealing with food.

Certain factors make digestive disorders more likely. Knowing what these factors are and what you can do about them can help you to avoid some common gut problems and spot food and drink-related hazards.

UNMODIFIABLE RISK FACTORS

As with heart disease, some digestive disorders tend to run in families. These include colon cancer, inflammatory bowel disease, and celiac disease. About 20 to 25 percent of colon cancer cases are caused by inherited abnormalities in genes. Having a parent with one of these conditions does not mean you'll definitely develop it as well, but you are more likely to.

Although you obviously have no control over your genetic background, you can still take effective action to reduce your risk of digestive disorders: You can avoid lifestyle behaviors known to contribute to triggering a disease; you can adopt a preventative approach through diet; and your doctor can enroll you in a regular screening program to detect any disease at an early and much more treatable stage. For example, if you are at risk of colon cancer because of a strong family history of the disease, you can have regular screening by a procedure called colonoscopy (see pages 48–49 for more details on screening). Age is a contributing factor in some diseases: Colon cancer and diverticular disease, for example, both become more common after the age of 50.

MODIFIABLE RISK FACTORS

Although you can't control your age and inheritance, there are other factors that you can influence, which can make a real difference to your future digestive health.

- **Stop smoking** Smoking increases the risk of a range of digestive complaints—heartburn, esophageal and mouth cancer, peptic ulcers, chronic diarrhea, and Crohn's disease. Every year, smoking causes more than 13,000 deaths from cancer of the esophagus, stomach, throat, and pancreas. If you smoke cigars, you still run the risk of developing cancers, particularly of the mouth, throat, and esophagus.
- **Limit alcohol consumption** People who regularly drink more than the recommended limit (1 ounce a day for women; 2 ounces for men) are at risk of developing gastritis (inflammation of the stomach lining, peptic ulcers, and cirrhosis of the liver. Heavy drinkers are also more likely to develop cancer of the digestive tract.
- **Cut fat intake** Researchers estimate that more than one third of cancers may be related to eating habits. Diets high in fat have been linked to increased risk for various cancers, including colon cancer.

The main culprits—food-poisoning microorganisms

Many different microorganisms cause food poisoning. Some are ubiquitous; others tend to be found in particular types of food. The timing between eating a contaminated food and the onset and duration of symptoms varies widely, depending on the particular organism responsible. Check out the chart below for the most common culprits.

Bacterium	Symptoms	Common foods	Time from eating	Duration of symptoms
SALMONELLA	Diarrhea, vomiting, fever, headache, abdominal cramps	Raw meat, poultry, eggs, and egg products	12–24 hours	1–7 days
STAPHYLOCOCCUS AUREUS	Diarrhea, vomiting, nausea, abdominal cramps	Cold meat, poultry, custard, and cream	2–6 hours	6–24 hours
CLOSTRIDIUM PERFRINGENS	Diarrhea, abdominal cramps	Cooked meat, poultry, fish, stews, pies, and gravy	8–22 hours	24–48 hours
BACILLUS CEREUS	Vomiting, diarrhea, abdominal cramps	Boiled or fried rice that has been kept warm or inadequately reheated	1–16 hours	12–24 hours
ESCHERICHIA COLI (E. COLI)	Diarrhea, vomiting, mild fever, abdominal cramps	Many raw foods	12–72 hours	1–7 days
CAMPYLOBACTER JEJUNI	Diarrhea, abdominal cramps, fever, sometimes bloody stools	Milk and raw poultry	3–5 days	Days to weeks
VIBRIO PARAHAEMOLYTICUS	Mild abdominal cramps, diarrhea	Seafood and fish	12–18 hours	2–5 days
LISTERIA	Mild fever. In pregnancy, can result in miscarriage or stillbirth	Unpasteurized milk and dairy products, cold meats	2–49 days	Days to weeks

Current guidelines recommend a total fat intake of no more than 30 percent of daily calories, which converts to about 95 grams of fat for a man and 75 grams for a woman.

- Eat plenty of fresh fruit and vegetables A diet rich in fruit and vegetables seems to protect against cancer of the esophagus, stomach, and colon. Eating a high-fiber diet also reduces the risk of diverticular disease—a sometimes painful condition that affects the large intestine. Some fruit and vegetables may help more than others: Research suggests that oranges, spinach, corn, strawberries, bananas, and apples are especially good because they are rich in antioxidants, chemicals that neutralize cancer-causing free radicals. But don't worry about matching a specific disease, just focus on eating five or more servings a day from a variety of fresh produce—leafy greens, carrots, broccoli, tomatoes, garlic, onions, and fruit of all kinds.

Other risk factors for gut health are stress (see pages 52–53), caffeine (see page 60), and lack of exercise (see pages 84–89).

A GI HEALTH RISK?

Food is a common source of gastrointestinal illness. Bacterial contamination of food is usually to blame for acute gastroenteritis (see chart, page 43), but viruses can also be the cause. It can take a few hours or several days before symptoms of food poisoning develop, so pinning down the offending food may be difficult.

The time elapsed between eating infected food and getting ill varies according to a number of factors: the organism responsible, the number ingested, and the presence of toxins—poisonous substances produced by certain bacteria. A person's age and general health also come into the equation. If vomiting starts within an hour of eating a food, the cause is usually a bacterial toxin or, rarely, chemical poisoning from metals.

A matter of numbers

It has been estimated that that 76 million Americans get food poisoning each year. Of that number, 300,000 are hospitalized and 5,000 die from these foodborne illnesses. Infants, the elderly, and the those with compromised immune systems are most susceptible. The U.S. Centers for Disease Control announced recently, however, that in the past five years, food poisoning from dangeous bacteria such as *E. coli* and *Salmonella* has declined substantially, suggesting that measures taken to make food supplies safer are working.

In the period 1996 through 2001, the rate of poisoning by *E. coli* fell 21 percent, *Salmonella* 15 percent, and *Listeria* 35 percent. *Campylobacter* dropped at least 35 percent. Only the rate for *Vibrio*, a bacteria that is found in raw oysters, rose 83 percent.

Health officials attribute the declines to better regulation throughout the U.S. food industry, including more stringent inspections at slaughterhouses and in the seafood industry. Because of the new regulations, tens of thousands of food poisoning infections are prevented each year. In 1997, the government began putting less emphasis on spot checks and instead required seafood plants to show proof of steps taken to prevent contamination. The regulations were soon expanded to cover meat production. Then, egg refrigeration requirements were imposed on supermarkets and restaurants, and fruit and vegetable juices and imported foods were similarly regulated.

American consumers are also credited with contributing to the decline in incidences of food poisoning. It appears that Americans are paying more attention to food safety, perhaps cooking meats and eggs more thoroughly, defrosting foods appropriately, and washing their hands before eating.

Food poisoning in other countries, however, for example the United Kingdom, has continued to escalate in recent years. The finger of blame has been pointed squarely at modern intensive farming methods, which allow the spread of bacteria among animals and encourage drug-resistant strains by the routine use of antibiotics. Battery-reared poultry are also under prolonged stress, which is believed to lower their immunity to infection.

Americans cannot rest on their laurels, however. It has been estimated that medical costs and lost wages from poisoning by foodborne *Salmonella* alone is more than $1 billion per year. But, they are on the right track.

The multiplication factor

Bacteria multiply at an astounding rate: Each bacterium may replicate every 10 to 30 minutes, so a single organism may become many millions within just a few hours. Bacteria need a food supply (almost any will do, but meat, milk, eggs, poultry, fish, seafood, and products made from these are favorites), warmth (a temperature between 68°F and 98.6°F), moisture, air, and sufficient time.

You never know if bacteria have grown on food, because there may be no mold, smell, or change in taste. So don't be tempted to eat food that is way past its use-by date because it still looks and smells okay. You could be putting your digestive

TALKING POINT

Could "clean" intestines cause inflammatory bowel diseases?

More people than ever before have inflammatory bowel diseases such as Crohn's disease. These diseases are thought to result from the immune system attacking proteins in cells of the large intestine. One theory holds that immune system development is linked with our early exposure to micro-organisms. With improved sanitation we now meet fewer micro-organisms in childhood and the immune system overreacts to harmless organisms.

SIMPLE STEPS TO SAFE EATING

Some aspects of food hygiene are common sense, but others may be new to you. Look over these simple measures to limit the spread of bacteria in your kitchen, which can make the difference between an upset-free digestive system or one in persistent trouble.

PERSONAL HYGIENE

- *Wash your hands before touching food. If you handle raw meat, fish, or poultry, wash you hands again before touching any other food. Always wash hands after handling waste food and after visits to the toilet.*
- *Dry your hands thoroughly, because dry hands have fewer bacteria.*
- *Any cut, burn, or skin break on the hands—no matter how small—may contain additional bacteria and should be covered with a bandage.*

PREPARING AND STORING FOOD

- *Kitchen surfaces and utensils should be cleaned thoroughly and frequently with hot water and cleanser.*
- *Use separate chopping boards for raw and cooked meat and vegetables.*
- *Salad, vegetables, and fruits should be well washed or peeled.*
- *Wash all surfaces and utensils with cleanser after contact with raw food, especially meat, and before using them for anything else.*
- *Ideally, meat and fish should be cooked or frozen the day of purchase.*
- *Frozen food should be thawed properly in the refrigerator (not left out on the counter) before cooking. Once thawed, the food should be cooked and eaten within a short time.*
- *Be particularly vigilant when dealing with high-risk foods, such as meat, eggs, poultry, and dairy products; they all require extra care.*

SAFE COOKING

- *The safest methods of cooking are pressure-cooking, grilling, roasting, or frying because of the high temperatures that are reached.*
- *To destroy potentially hazardous bacteria, cooking must be thorough; if you are roasting a large joint of meat, use a meat thermometer to check that it has reached the right temperature.*
- *Avoid "part-cooking" in advance. This can create the ideal conditions for bacterial growth. It is safer to keep raw meat in the refrigerator overnight and then cook it the next day.*
- *If you can't eat food when planned, cook it while the food is fresh, then freeze it, rather than freezing the raw food and cooking it at a later date.*
- *Once cooked, either eat food right away or cool it rapidly and refrigerate it within 1½ hours to avoid the multiplication of any surviving bacteria or heat-resistant spores.*

system at risk. Traditionally, people pickled foods or stored them in salty water to preserve them and discourage bacterial growth. These extreme conditions—acidic vinegar or high salt concentrations—prevent bacteria from multiplying. Other sterilization techniques involve extremes of temperature. High temperatures (above 145°F for at least 30 minutes, as in the pasteurization of milk) or low (refrigeration at 33–41°F, which slows bacterial growth, or freezing, which halts bacterial growth but doesn't kill bacteria) also prevent any possible bacterial colonies from growing.

Eating out? Spot the danger signs

At home, eating is under your control, but what about eating out? Look for obvious signs of poor hygiene in restaurants, such as unclean surfaces and poorly washed crockery and cutlery. Beware of food served less than piping hot. If you are at all concerned about hygiene in a restaurant, either leave immediately or avoid dishes that contain poultry, meat, fish, or seafood.

TRAVEL HEALTH

Today, more people than ever before travel to worldwide destinations for work or pleasure. With travel comes the need for awareness about diseases that might be contracted abroad and advice on how they can be avoided.

Travelers' diarrhea

Every year more than a million people have diarrhea while on vacation. It usually lasts no more than four days and is caused by bacteria in the locale that are new to your stomach, although harmless to the locals. Often, travelers' diarrhea is caused simply by a change in diet, which disturbs your GI flora, or from eating too much fruit or drinking too much alcohol. But some cases are caused by intestinal infections. The most common culprit is a toxin released from the bacteria *E. coli*. This toxin causes stomach cramps, vomiting, and watery diarrhea. Dysentery, a particularly severe form of diarrhea, comes from ingesting food or water contaminated with *Salmonella, Campylobacter,* or *Shigella* bacteria. Other causes are rotavirus, *Giardia lamblia*, and *Entamoeba histolytica*, but one quarter of cases have no identifiable cause.

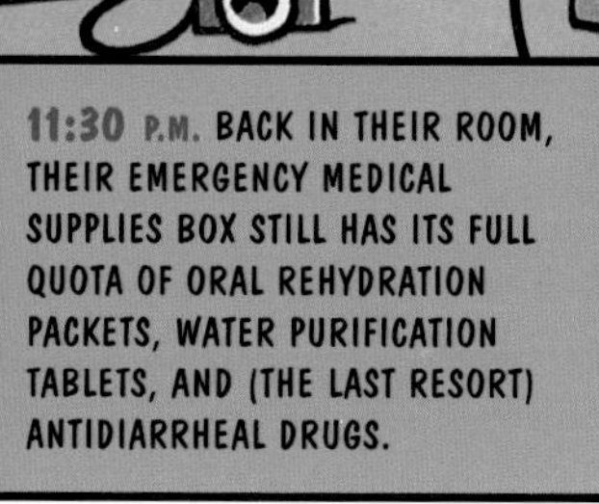

So what should you do if you develop diarrhea? Your main concern should be to avoid becoming dehydrated, so drink plenty of fluids. Oral rehydration salts are best—you can take some with you, just in case—but add them to bottled or boiled water. By combining salt and sugar in a specific ratio, these rehydration salts maximize the water absorbed in your large intestine to restore your hydration status (see page 120 for a simple recipe).

While recovering, stick to a bland diet—plain rice, bread, and soup are good. If you don't have time to wait for the rehydration salts to take effect, you might consider using antidiarrheal drugs to control the symptoms. They are useful if you have to travel but may prolong the diarrhea by retaining organisms in the GI system. They can also cause constipation. Consult a doctor if your diarrhea lasts for more than 12 hours, is blood-stained, or if you have a high fever.

Avoiding tummy troubles

There are several positive steps you can take to avoid "the runs" on vacation.

- When eating out, choose places used by travelers as well as locals, and stick to those that are busy.
- Fast-food stands are a matter of judgment—check that a stand seems hygienic and that meat is kept cool and cooked thoroughly.
- Ice cream and dairy products such as yogurt are best avoided in developing countries.
- Travel guides usually say whether tap water is safe to drink, but water that is safe for locals may contain bacteria that upsets your GI system simply because it is unfamiliar. If in doubt, buy bottled water (with the seal intact), boil water, or use purification tablets.
- If you are not drinking the local water, try to avoid ice cubes in drinks because these are often made with tap water. Avoid salads, which will probably have been washed in tap water, and fruit, unless it can be peeled.
- Don't assume that it is safe to eat and drink everything in your hotel because it has a four- or five-star rating. Stick to the advice given earlier to be sure.

Can you take drugs to prevent diarrhea on vacation?

Most people taking a two-week vacation in the Mediterranean will find antidiarrheal drugs completely unnecessary. But, if you're traveling to a remote region where it's unlikely you'll find medical help, you may want to consider talking to your doctor about taking along a prescription of antibiotics, such as ciprofloxacin, to prevent diarrhea. This can be taken beforehand or as soon as symptoms occur.

ASK THE EXPERT

Be involved in your health care

Your health is your most vital asset, and by working in partnership with your physician, you can help to maintain your digestive health. Your doctor can give you advice on how to prevent and pick up on any problems early.

Salad days

Living a GI-friendly lifestyle is not just about keeping to a healthy diet—it also includes learning when to visit your doctor and when to deal with things yourself.

CHECK OUT ANY SYMPTOMS

All of us suffer from constipation, diarrhea, or excessive bloating from time to time. Normally, if we think about it, we can put it down to a few days of an unusually low-fiber or high-fiber diet. Short-lived symptoms are usually nothing to be concerned about. But there are some symptoms that you should take seriously and consult your doctor about.

- **Abdominal pain** This can be associated with a multitude of conditions. The severity of the pain doesn't always reflect the severity of the condition. Severe abdominal pain can be caused by something as innocuous as gas or the cramping of viral gastroenteritis, while relatively mild pain may be a symptom of early appendicitis. Pain that is persistent should alert you to consult your doctor.
- **Indigestion** Although common and usually harmless, indigestion that continues for more than one to two weeks, especially in people over 50, may be a sign of a more serious condition such as esophageal or stomach cancer.
- **Difficulty or pain swallowing** Although this might be caused by simple inflammation of the esophagus, this is an important and serious symptom. Always consult your doctor if it persists for more than a few days.
- **Change in bowel habit** If you have diarrhea or constipation that continues for more than five days, see your doctor. You may have an obstruction or other significant problem.

Use the "Help your doctor to help you" box (opposite) to think about questions your doctor may ask you.

SCREENING FOR ALL?

Your physician can advise you on any tests that might be useful, especially as you get older. Many doctors run regular clinics, such as well-woman and well-man clinics, at which you can have a general medical check-up—the aim being to pick up problems before they become serious or difficult to treat successfully. The frequency of screening tests depends on your previous health, your family history of digestive problems, and your present state of health.

The 50-year milestone

From the age of 50, the risk of digestive problems, such as colon cancer, increases, and you might want to talk to your doctor about a check-up or screening test.

Your doctor can get a lot of information from doing a general examination of the digestive system, and if you have any digestive problems, this will be the first step in finding out what's wrong. This examination usually includes looking at your physical appearance: the hands, eyes, face, and abdomen. Your doctor may also carry out a physical examination of your rectum, which involves placing a gloved finger inside to feel for any abnormalities.

Your doctor may also arrange blood tests to look for anemia and to check liver function, especially if you drink alcohol to excess.

Screening for colon cancer

The emerging consensus is that everyone should be screened for colon cancer from the age of 50, although this is not done routinely in the U.S. General annual screening requires a simple test to detect tiny traces of blood in the stools—the fecal occult blood test (FOBT). A positive result will lead to more precise, but invasive, investigations, such as a colonoscopy, in which a doctor looks inside the colon with a special instrument (see page 115), flexible sigmoidoscopy (see page 114), and barium enema (see page 108).

About 150,000 people in the United States will be diagnosed with colon cancer this year; 9 out of 10 people treated at an early stage survive for at least 5 years.

In general, doctors recommend a flexible sigmoidoscopy every 5 years. A colonoscopy is recommended every 10 years, and a double-contrast barium enema, every 5–10 years. Treatment for colon cancer has a much higher success rate if it is detected and treated early.

Family fortunes

Some people are at higher risk of colon cancer than others. If someone in your family has colon cancer, you may need to be tested at regular intervals before you reach 50 (see page 97). If a parent or sibling has had colon cancer (especially at a fairly young age), doctors often advise people to have a colonoscopy once every 5 years, starting 10 years before the age the relative was diagnosed. People who have conditions that affect the large intestine, such as ulcerative colitis or Crohn's disease, may also need to be screened more often than the general population because they are at higher risk of developing colon cancer. Doctors recommend that people with extensive ulcerative colitis have a colonoscopic investigation every two to three years, starting about ten years after their initial symptoms.

A question of taste
Papillae (red) surround a central taste pore (blue), which leads to a barrel-shaped taste bud underneath. Taste bud numbers decline as we get older.

HELP YOUR DOCTOR TO HELP YOU

Is it serious?

Any of the following symptoms warrant a prompt visit to the doctor. Before the appointment, it may help to write down details of when symptoms started, their severity, or their frequency.

- *Prolonged diarrhea. When did it start? Do you have any other symptoms?*
- *Bleeding from the rectum, bloody stools, or vomiting blood. When did it start?*
- *Persistent abdominal pain. Is it sharp, crampy, or dull?*
- *Unexplained loss of appetite and weight loss*
- *Change in bowel habits, such as persistent or alternating diarrhea or constipation, especially if you are over 50*
- *Yellow discoloration of your skin or whites of the eye*

THE AGING GI SYSTEM

As we get older, subtle changes take place in our digestive systems. We may feel inclined to add extra salt or spices to food, for example, to compensate for a natural decline in

Absorbing medicines

Medicines are usually swallowed and absorbed in the GI system. Some drugs are absorbed directly from the stomach, but most are absorbed in the small intestine. The rate of absorption depends on whether or not there is food in the stomach and what form of drug is taken. Prescribed medicines usually come with instructions to take them before, with, or after food. Some medicines don't damage the stomach mucus lining and are best taken before a meal. Others can irritate the stomach, making you feel nauseous; they must be taken with or after food, which reduces their irritant effects. Soluble drugs travel through the intestinal lining quickly and are absorbed rapidly. Next in terms of speed are suspensions, followed by tablets, which have to be broken down before the drug can be used.

the number of taste buds. The tongue can also help us to spot dietary problems: Vitamin B_{12} deficiency may cause it to go pale, whereas a shiny tongue may reflect a lack of iron, B vitamins, or folic acid.

A dry mouth

A common complaint among older people is dryness of the mouth, sometimes associated with painful small ulcers. If you have this problem, check with your doctor, because medication could be to blame. Other causes include diabetes, infections, or skin diseases. Chewing gum can stimulate saliva production.

Slowing down

The production of stomach acid and digestive enzymes often declines with age, leading to poor absorption of nutrients from food into the body. There is a gradual sluggishness of the bowels, and specific bowel disorders such as diverticular disease, are more common in older people.

MEDICINES AND THE GI SYSTEM

Some medicines can have damaging effects on your digestive tract, especially when taken every day. If you take medicines regularly and often suffer from digestive complaints, alert your doctor so that you can discuss available alternatives.

Aspirin and related drugs

Nonsteroidal antiinflammatory drugs (NSAIDs), which include aspirin and ibuprofen, are prescribed for arthritis and other inflammatory conditions but can also be bought over the counter as painkillers. When taken regularly, these drugs may cause stomach ulcers. They have a direct irritant effect on the stomach lining and also inhibit formation of chemicals known as prostaglandins, which are responsible for maintaining the protective mucus lining in the stomach. With prolonged NSAID treatment, normal stomach acid may dissolve the weakened mucus and cause ulcers.

The problem is worrisome because millions of people use NSAIDs regularly to treat everything from headaches to kidney stones. In most cases of peptic ulcers caused by such medications, however, the ulcers disappear when NSAID use is discontinued.

Try to take NSAIDs with food to minimize the irritant effect on the stomach. Also, ask your doctor for the "GI-friendlier" form.

Other drugs

According to one study, 30 percent of over-60-year-olds take laxatives, probably in the mistaken belief that a daily bowel movement is necessary for health. Long-term use of laxatives, particularly those derived from senna pods, may cause degeneration of the nerves supplying the colon, making the constipation worse.

About 6,000 people die of ulcer-related complications in the United States each year.

Antacids are taken to ease indigestion and heartburn. Long-term use can cause diarrhea or constipation, but occasional use should not cause problems.

Antibiotics may adversely affect the natural population of bacteria in the large intestine—the digestive flora—and diarrhea is a common side effect of antibiotic treatment.

HEALTHY LIVING FOR YOUR DIGESTIVE SYSTEM

The simple fact that digestive illness varies so dramatically around the world suggests that lifestyle is a key element in maintaining digestive health. Factors that have the potential to wreak havoc with your finely tuned digestive system are explored in the following pages, together with the positive ways in which you can take your digestive health into your own hands and give yourself a real appetite for life.

Combating stress

Long-term stress can be particularly harmful to your digestive system, contributing to stomachaches, diarrhea, and irritable bowel syndrome. For some of these problems, the remedy lies in learning to relax and keep the pressure at bay.

A GUT REACTION

Stress affects us all to a greater or lesser extent, be it through work, through a busy lifestyle, or even from being bored. Physical and emotional stress switch the body to a heightened state of alertness to get ready to fight or else make a run for it—but at the expense of digestion.

At times of stress, blood supply to the GI system is virtually shut down and digestion is sacrificed in favor of faster heart rate and breathing. Consequently, the digestive system is deprived of essential blood and nutrients, and digestion is temporarily brought to a halt. Saliva production falls, resulting in a characteristic dry mouth.

Curbing hunger

Diversion of blood from the GI system can also affect appetite. Reduced blood flow is thought to slow or even stop the stomach from emptying completely. This may make us feel either not hungry or nauseous. At the same time, the body starts to mobilize reserves of fuel (in the form of the carbohydrate glycogen) that have been stored in the liver, converting it to glucose and then transporting it to where it's needed most. This glucose may also directly suppress the brain's appetite center. Once the stress is gone, hunger and digestion restart and continue.

The litmus test

Exposure to stressful stimuli may alter the secretion of gastric acid and pepsinogen, an enzyme that stimulates acid production. As a result, the pH of the stomach falls—becoming more acidic—and secretion of the protective mucus layer of the stomach itself is reduced. At the same time, reduced blood supply to the GI system and excessive production of the chemical prostaglandin can cause erosions in the stomach wall, and small areas may be worn away. This reaction can occur following extreme stress, such as after automobile accidents, and may aggravate peptic ulcer disease.

In the long term

Though these changes are beneficial for coping with serious immediate "threats," repeated or prolonged stress may be harmful in many ways.

Long-term psychological stress has been linked with immune system problems, for example, which increases the risk of acute illnesses such as viral infections. At the lower end of the digestive system, the stress response can make the large intestines sluggish. If the stress is prolonged, this reaction can be a be a major cause of constipation.

The rewards of relaxation
The abdominal breathing used in many Eastern techniques helps to exercise the diaphragm and relax both body and mind.

Mint—a natural remedy
With its natural antispasmodic properties, mint is a great remedy for digestive complaints. Infused in boiling water to make tea, it can be helpful for people with IBS.

IRRITABLE BOWELS

Stress is one of the many factors that have been linked to irritable bowel syndrome (IBS). IBS is very common and affects more women than men. It causes a number of symptoms, such as lower abdominal pain, bloating, and episodes of constipation or diarrhea. So how can stress have such a dramatic effect?

Some studies have shown that individuals with IBS have overly sensitive muscles in the wall of the large intestine. These muscles react excessively to nerve impulses from the brain when under stress. These nerves control the intestinal muscles that squeeze and relax, which propels food along its journey.

This hypersensitivity to the stress response may account for some of the feelings of bloating, pain, or altered bowel habit in people with IBS. What's more, a vicious circle can develop: While the intestinal muscles are overstimulated, pain messages are relayed from sensory nerve endings in the digestive tract to the brain, and the brain responds with the classic stress response, stimulating the hyperactive intestines even further.

CALM DOWN YOUR BOWELS

Techniques to relieve stress can help to alleviate some of the symptoms of IBS in some people. If you suffer from IBS, try this simple relaxation technique. Find somewhere to sit or lie down quietly for 20 minutes. Begin by concentrating on your breathing: Breathe deeply, filling and emptying your lungs as completely as possible. Now concentrate on tensing and relaxing each part of your body one at a time. Start at the toes and work your way up to your eyebrows. Try doing this exercise every day.

De-stressing therapies

Some people benefit from more specific therapies, especially if they are also anxious or depressed. Some people benefit from psychotherapy or cognitive behavioral therapy, which attempts to modify behavior and your response to certain situations. Others find hypnosis helpful or biofeedback, where you learn to control bodily functions through direct feedback about them.

Massage can sometimes help to relieve cramping abdominal pains, especially in children. Slow, rhythmic, and firm massage strokes are applied in a clockwise circling motion around the perimeter of the tummy—following the outline of the large intestine. A hot-water bottle may provide some welcome pain relief.

Calming medicines

Antispasmodic medicines act on the wall of the GI system, reducing the transmission of nerve signals to the wall of the large intestines and thus helping to stop spasms.

Mint is a natural antispasmodic. Peppermint tea has long been used as an aid in digestion, and for those who can't stomach the tea, peppermint oil capsules deliver directly to the lower part of the large intestine for pain relief. See pages 76 to 77 for more remedies for digestive problems. If you have persistent pain or symptoms, such as a change in normal bowel habit, weight loss, or vomiting, it is important to seek medical advice.

IT'S NOT TRUE!

"Stress is the main cause of ulcers"

It's a widely held belief that stress causes peptic ulcers, but research has shown that there are other more important factors at work. Infection with *Helicobacter pylori* bacteria, smoking, and taking aspirin or other nonsteroidal antiinflammatory drugs for a prolonged time all increase the risk of peptic ulcers. Although day-to-day stress may result in a small increase in digestive acidity—and consequent feelings of discomfort or indigestion—it doesn't cause actual ulceration.

Alcohol: Ally or culprit?

Is alcohol a friend to your digestive system? After all, a moderate alcohol intake has been shown to be good for your heart. Or is every alcoholic drink simply a bad influence on your digestive health?

Much folklore surrounds the drinking of alcohol and its impact on digestion. All sorts of alcoholic drinks have been claimed to aid digestion, but there is little evidence of this.

AN ALCOHOLIC HISTORY

The idea of the apéritif, the before-dinner drink, came from Sweden, when in the 18th century a pre-dinner snack, called a smörgasbord, consisting of herring, cheese, bread, and aquavit (an alcoholic spirit made from potatoes) was served. The idea spread to France, and strong spirits were served before dinner to stimulate the appetite.

Spirits were believed to have medicinal properties long before they were drunk for pleasure. As far back as biblical times, people valued wine as a medicine, believing it kept the digestion working well and protected the body from bad "humors" and cleared "maggots from the brain." Bénédictine liqueur, for example, was originally produced as a way of preserving herbs (growing around the Benedictine Abbey in Normandy) for an elixir to combat local diseases.

AN ABSORBING TALE

Once you've swallowed a mouthful of an alcoholic drink, it flows down the esophagus through a sphincter (an opening controlled by a band of muscle) and into the stomach. This is where absorption into the bloodstream begins. Up to 10 percent of the alcohol you've drunk will be absorbed through the mucus-lined stomach wall. The rest passes into the small intestine—the main site of alcohol absorption.

Lining the stomach

The contents of your stomach influence what happens next in alcohol's journey. If you've eaten recently, the pyloric sphincter between the stomach and the small intestine will be closed. This shut sphincter keeps food in the stomach long enough to be broken down by acids released through the stomach wall. It also keeps alcohol from passing straight through to the small intestine and so it doesn't get absorbed so quickly.

The rule is, then, that if you line your stomach before drinking, you are less likely to end up under the table after a few drinks. Instead, your blood alcohol level will reach a slower, lower peak. If you drink on an empty stomach, your blood

Wine-loving nations

On average, Americans drink only 8 quarts of wine each year. People in the U.K. drink 15 quarts, and the French drink 63 quarts per year.

BEFORE AND WHILE DRINKING

Prevention, plus common sense, is better than cure: Use these tips to avoid a hangover.

- Drink a pint of milk to line your stomach before you begin drinking.
- A carbohydrate-rich snack helps to slow down alcohol absorption.
- Intersperse alcoholic drinks with mineral water or fruit juice.
- Sip your drink slowly, and try not to drink more than your limit.

HANGOVER CURES

If you've overdone it with alcohol, start to remedy the situation as soon as possible for maximum effect.

Drink a large glass of water to replenish the body's fluid levels.

Have a large mug of sugary tea or honey mixed in a glass of fruit juice.

Try scrambled eggs: they contain elements that can mop up the toxins.

Nonaspirin painkillers are best for a hangover headache.

alcohol level will reach a peak level after only one hour. But if you've eaten recently, especially if you've had a fatty meal, the absorption of alcohol into the bloodstream is much slower—peak alcohol levels aren't reached for three to six hours, and the levels may be reduced by 25 percent.

On an empty stomach, alcohol passes straight through the stomach into the small intestine, where it is absorbed quickly through the millions of tiny fingerlike projections called villi. Alcohol is then transported by the bloodstream directly to the liver and from there to the rest of the body.

Breaking down the poison

The body sees alcohol as a poison, so it does its best to get rid of it as quickly as possible. The first step in the detoxifying process is under the control of an enzyme called alcohol dehydrogenase, which is present in the stomach wall and in the liver. Another liver enzyme, aldehyde dehydrogenase, completes the conversion of alcohol into harmless constituents. The liver contains 17 varieties of aldehyde dehydrogenase, and within an hour it can handle ½ fluid ounce of pure alcohol.

We don't all have the same levels of these enzymes, however. In fact, some races have very little aldehyde dehydrogenase and are particularly vulnerable to the effects of alcohol. Half the population of Japan, for instance, doesn't have enough effective aldehyde dehydrogenase enzyme—just one or two drinks lead to the nausea and headaches normally associated with a bad hangover.

Regular drinkers also seem to metabolize alcohol faster. This is the result of another set of detoxifying enzymes being activated in regular drinkers. As a result, a lower peak alcohol level is achieved than in an average drinker. This is called tolerance and accounts for the fact that regular drinkers can drink more before becoming drunk.

Why do people have an apéritif before eating?

It's a commonly held view that a small drink before a meal—a glass of dry white wine or sherry—may stimulate the appetite. Most traditional apéritifs are slightly bitter and contain a relatively low percentage of alcohol. Apéritifs supposedly "open" the digestive system and stimulate the appetite before a meal. The small amount of alcohol gets acid production started so that the stomach is ready to digest food right away. However, there's no real evidence that one particular type of alcoholic drink is any better than another as an appetite stimulant.

ASK THE EXPERT

WHAT'S YOUR POISON?

Another factor in the absorption equation is alcohol concentration. Drinks with a low alcohol concentration, such as beer, are absorbed more slowly than high percentage liquor. This has to do with the concentration gradient between one side of the small intestine and the other side. If the alcohol concentration in the lumen of the small intestine is 10 percent, the alcohol passes through the walls at a slower rate than if the alcohol concentration were 40 percent. In

The strongest known alcoholic drink was an Estonian spirit, distilled from potatoes, which measured a staggering 98 percent on the alcohol content scale. Production of this spirit ended in 1978.

How strong is your poison?
Drinks with a high alcohol content (ABV—alcohol by volume) are more rapidly absorbed through the small intestine than low-alcohol drinks. You can use this list to see the alcoholic content of your favorite drink.

addition, the greater volume of liquid present when drinking beer slows the emptying of the stomach. Drinks with high alcohol concentration also irritate the stomach's lining, causing secretion of mucus and a slowed emptying of the stomach. Carbonated drinks, such as champagne, speed up the emptying of the stomach into the small intestine, where alcohol is quickly absorbed.

WOMEN VERSUS MEN

Women are more susceptible to the effects of alcohol than men for a number of reasons. One difference is they have a lower level of alcohol dehydrogenase, which breaks down alcohol, so more is absorbed into the bloodstream. But there are other differences, too. In women, water comprises a lower percentage of total body weight than it does in men of the same weight. So, if a man and a woman both drink three glasses of wine, women will have a higher percentage of alcohol in their bodies.

FAT OR THIN?

Broadly speaking, the heavier you are, the greater the volume of water in your body. More water means that any alcohol consumed is more diluted in the blood and body tissues. This would seem to give the overweight person an advantage when it comes to dealing with alcohol, but this advantage is lost if the excess weight is fat rather than muscle. Fat cells contain less water than muscle cells, so the body's water volume is correspondingly less in a person with more fat and less muscle.

PERIODIC DRINKING

Alcohol absorption is affected by the menstrual cycle. Many women find they succumb to the effects of alcohol faster in the couple of days around ovulation and in the time leading up to menstruation. This coincides with low levels of progesterone, which tends to speed the emptying of the stomach. When progesterone dominates (after ovulation), the stomach empties more slowly, so there is more time for alcohol to be broken down.

THE WEIGHT CONNECTION

All but 5 percent of the alcohol you consume at one time is broken down in the liver. The rest is eliminated in the breath, sweat, saliva, and urine. A side product of alcohol breakdown in the liver is heat—in other words, calories—for the body to use as energy. Alcohol has a Caloric value of 7 per gram. These calories are often referred to as "empty calories," because they contain no nutrients. Someone drinking large quantities of alcohol and eating a full diet puts on weight because of the high sugar content of most alcoholic beverages.

People who regularly abuse alcohol are three and a half times more likely to die in a given time than people in the general population.

THE LONG-TERM VIEW

In past centuries, fermented alcoholic drinks were about the safest form of liquid intake—the fermentation process got rid of otherwise harmful bacteria in contaminated water. Now, other health reasons are used as an excuse to drink: namely, that it may be good for the heart. But when it comes to the digestive system, alcohol does nothing but harm.

GI system–damaging effects

Research has shown alcohol to be one of the worst stomach irritants we can ingest. It increases the production of stomach acid, upsetting the balance of acid to mucin—the protective substance lining your stomach walls.

The effects of this imbalance vary from mild heartburn or nausea to a condition called gastritis, in which the stomach actually begins to digest its own lining. Any alcohol over 40 percent ABV directly erodes the stomach mucosa, but all forms of alcohol can inflict similar wear and tear on the body if you drink enough.

In addition to damage to the stomach, alcohol also causes inflammation and ulcers in the intestines—especially in the first part, the duodenum.

Wreaking havoc with the liver

In the United States, liver disease is commonly caused by alcohol misuse. The first sign of liver damage is the deposition of fat in the liver, which can occur within days of heavy drinking. Abstinence from alcohol reverses this stage. If a person carries on drinking, fibrous scarring ensues, followed by irreversible damage. Cirrhosis may develop over a short period of time or, despite drinking sizeable amounts over a lifetime, may never develop.

The cancer risk

Experts believe there is a link between alcohol consumption and cancers of the digestive system, breast, throat, and liver. This link particularly holds true for high-alcohol drinks such as hard liquor, compared with beer or wine. Researchers believe this to be attributable to the higher content of nitrogenous chemicals in hard liquor; therefore, drinking them with mixers will not reduce the risk.

Drugs and digestion

Most drugs, whether prescribed, bought over the counter, or taken recreationally, have some effect on the gastrointestinal tract. The most common side effects are nausea or diarrhea. Some recreational drugs have more specific effects.

Marijuana

This is the most commonly used recreational drug in the U.S. It creates a mood of relaxation, increased sociability, and awareness of colors and sounds. It can cause hallucinations, nausea, and vomiting. It also tends to cause a dry mouth and an increased appetite, known as the "munchies." Marijuana is currently illegal in the U.S.

Amphetamines (speed)

These drugs were developed during the 1920s as a stimulant for soldiers to prolong their periods of wakefulness and activity. Subsequently, they were used in the 1950s and 1960s both for their antidepressant properties and for their ability to suppress appetite and cause weight loss. The appetite-suppressant effect is shortlived, and once the effects of the drug have worn off, the person is likely to feel extremely hungry.

Ecstasy (3,4 methylene-dioxymethamphetamine)

The amphetamine derivative Ecstasy is popularly believed to be a safe recreational drug. After an initial feeling of stomach butterflies, a state of calm euphoria develops for about six hours. During this time, the appetite is suppressed and there is usually a sensation of a dry mouth. Once the effect has worn off, there is usually a state of increased hunger and tiredness as the body "catches up." There have been several deaths related to Ecstasy usage, usually from dehydration and overheating.

Tobacco and tummy trouble

Most people know about the big health risks from smoking—heart disease and lung cancer. But do you know how tobacco affects your digestive system? The nicotine that gets you hooked also has some diverse effects on your gut.

Since tobacco was first exported from America more than 500 years ago, billions of people around the world have spent a life hooked on one of nature's most addictive drugs. Within a week of smoking, an individual can become both tolerant (needing more each time for the same "hit") and dependent on tobacco. If the effect is so immediate, then it's no wonder that smokers find it so tough to quit.

Tobacco smoking is responsible for about 3 million deaths worldwide each year, yet people continue to smoke. The health-damaging effects of tobacco are well known and stem not just from nicotine but also from the hundreds of other chemicals that are inhaled when you smoke.

By the time the humble tobacco plant is processed into cigarette tobacco, it contains more than 400 different chemicals, some of which are poisons.

NICOTINE AND THE GUT

With each puff on a cigarette, 90 percent of the nicotine (plus other chemicals) remains in the lungs, whereas the remaining 10 percent quickly makes its way into the bloodstream and then around the body.

Nicotine itself has a number of effects on the GI system. Many smokers have distant memories of the first time they tried smoking: Physical sensations of nausea and dizziness are a common theme; some people are even physically sick. Although such noticeable effects of this gastric stimulation fade, inside the body of a smoker the nicotine is wreaking havoc with the digestive system. The long-term picture is worse: higher risks of digestive disorders such as ulcers and cancer of the mouth, esophagus, stomach and, possibly, the pancreas.

The ulcer problem

Many smokers particularly enjoy a cigarette after meals, claiming that it helps their digestion. But, once nicotine is in your system, your stomach suffers the consequences.

Does smoking a cigar really aid digestion?

Traditionally, cigars are smoked after dinner, but they do not contain any particular digestive agents. The real reason for the tradition has more to do with the time large cigars take to smoke—up to two hours—so that a person has longer to relax, which in turn may help digestion.

ASK THE EXPERT

Nicotine reduces blood flow in the stomach and slows down the speed at which the stomach empties into the small intestine. Nicotine also lowers the production of protective chemicals called prostaglandins, so the lining is more vulnerable to attack by gastric acid, leading to heartburn and indigestion. In the long term, smoking makes you more likely to develop peptic ulcers, and makes it less likely for ulcers you already have to heal. Twice as many smokers as nonsmokers have peptic ulcers.

Speedy gut syndrome

Nicotine's powerful chemical actions also target the nerves supplying the muscles of your digestive tract. This boosts the peristaltic transporting of food, which may sound like a good thing, but speeding up the transit time can lead to diarrhea.

The nicotine conundrum

Long-term smoking also has repercussions on inflammatory bowel diseases such as Crohn's disease and ulcerative colitis. Although some research shows that nicotine (in patch, not tobacco form) relieves ulcerative colitis; other studies have yet to confirm this. Smoking does, however, exacerbate Crohn's disease.

SMOKING AND SNACKING

Many people rely on the often-used excuse that when they stop smoking their appetites will run out of control, leading to inevitable weight gain. Nicotine is an appetite suppressant, and it probably works by prolonging the sensation of having eaten (satiety), which means that smokers are less inclined to snack or pick at food between meals. Once smokers quit, though, their appetites return, and some may start to snack more. Some individuals complain of constipation when they quit, which is probably related to the decrease in the stimulation of the intestinal muscle.

THE CIGAR CASE

Cigar smoking is often believed to be less dangerous than cigarette smoking because cigar smoke is not usually inhaled. However, there is a similar incidence of cancer of the mouth, throat, esophagus, and lungs among cigar smokers—attributable in part to the fact that many cigar users are ex-cigarette-smokers and still inhale.

EFFECTIVE STEPS TO QUIT

On average, people who have just stopped smoking eat 300 to 400 Calories more a day. But don't let this put you off—the benefits of quitting greatly outweigh the health problems from any small weight gain. Try these tips so that your no-smoking plan is also a no-weight-gain program.

- **Pick out the benefits** Make a list of all the health benefits of giving up smoking, keeping it in a prominent position; plan a date when you're going to stop and stick to it.
- **Enlist support** Encouragement from friends and family can make all the difference to whether you succeed or fail.
- **Snack healthily** Keep a stash of low-Calorie snacks (carrots or celery are ideal) to nibble on when you feel the urge to snack.
- **Load up on OJ** Try drinking plenty of orange juice. Oranges contain citric acid, which makes your urine acidic, too, and helps to speed nicotine removal from your body. The tangy taste should also make you less likely to crave a cigarette.
- **Drink lots of water** This can have a dual effect: flushing nicotine out of your body and preventing the constipation that is common in the few days after giving up.
- **Cut down on caffeine** Reduce the amount of coffee and tea you drink. There's some evidence that caffeine can make the nicotine craving worse.
- **Avoid contact with alcohol** For many people, alcohol and smoking go together, so it helps to stay away from bars for a while (this is also another way to reduce your Calorie intake).

Work it out of your system

Increase the amount of exercise you get. Not only will this make you feel great, but it will also boost your metabolism so your body burns off more of the Calories you consume. You'll also start to notice how much easier exercise becomes once you've stopped smoking.

Caffeine, the double-edged sword

The caffeine in your tea or coffee may give your brain a lift, but does it do the same for your digestive system? In moderation, it seems its effects on the GI system are mild and pose little cause for concern—but there are limits.

Caffeine stimulates your brain and boosts alertness, but it also has irritant properties when it comes to your digestive system. Caffeine's diuretic action (it makes you produce more urine) can leave you feeling dehydrated, which in turn may lead to a lack of water in the intestines and so to constipation.

Refreshing herbal tea
Instead of your usual cup of caffeine-loaded coffee or tea, try drinking an herbal tea, such as camomile tea, instead. Fruit teas are also caffeine-free and can be drunk hot or chilled.

STIMULATING THE GI SYSTEM

Caffeine boosts your metabolism slightly through stimulation of the autonomic nervous system, with a corresponding rise in heart rate and blood pressure. This metabolic increase means that food you eat is burned more quickly. Caffeine may also have a slight influence on suppressing the appetite.

Intestinal commotion

As well as making you produce more urine, caffeine also promotes more visits to the bathroom by stimulating the bowels. Even relatively moderate amounts of caffeine have a mild laxative effect—which can be a godsend if you're suffering from a bout of constipation.

Caffeine blocks receptors for adenosine—a signaling chemical that controls the balance between muscle contraction and relaxation. Adenosine promotes the relaxation of the intestinal muscles. When caffeine occupies adenosine's spot by blocking receptors, the smooth muscles of the intestinal wall contract more easily and fall into a characteristic peristaltic rhythm. This contraction and relaxation speeds the transit of intestinal contents through the GI system. Although this effect is well documented, no one has ever reported any harmful results from this mild laxative effect.

Paradoxically, if you're not a regular coffee drinker, caffeine can actually cause diarrhea.

Interfering with minerals

Caffeine can influence how certain substances are absorbed within the intestine. In particular, caffeine interferes with the absorption of iron. If you are taking iron supplements for any reason, don't slosh them down with a cup of coffee because you won't benefit at all. Caffeine can also reduce the absorption of other minerals, too, including calcium—important for bone formation—and zinc.

Babies lack the liver enzymes necessary to deal with caffeine, so breastfeeding women should limit their caffeine intake.

SENSIBLE CAFFEINE LIMITS

The average person in in the United States today drinks about three or four cups of tea or coffee per day– that's about 300 milligrams of caffeine. The major sources of caffeine are coffee (containing anywhere between 40 and 180 milligrams per cup), tea (about 50 milligrams per cup) and cola drinks (12–40 milligrams). The level of caffeine in diet drinks is the same as in standard cola drinks; so unless the label states that it's caffeine-free, assume it isn't.

How much caffeine do you think you drink on an average day? If the answer is more than five cups, then maybe it's time to consider cutting down.

Rather than suddenly stopping drinking caffeine altogether, try to reduce your caffeine intake by about half a cup of tea or coffee per day, substituting with another caffeine-free hot drink, such as herbal tea.

EATING FOR HEALTHY DIGESTION

We hear and read so many conflicting stories through the media about what we should or shouldn't eat that it can be difficult to know who or what to believe. Food is not the enemy of your GI system, however, and the overall message is really a simple one: A well-balanced diet that contains sufficient fiber can represent a potential fountain of good health, without compromising the great pleasure that food gives us. When problems do occur, adjustments to the diet or increased intake of certain "superfoods" can often provide the answer.

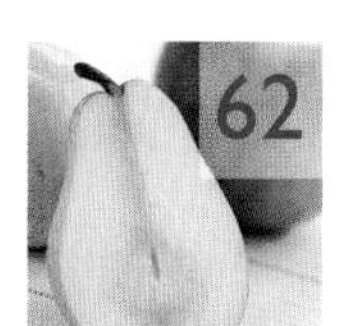

Scientists have not yet uncovered all the links between diet and health, but key components such as fiber are vital to your GI health.

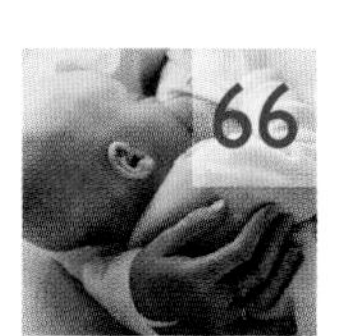

More than simply what you eat, GI health is about when and how you eat. We look at developing healthy habits from a young age.

Realistic and positive attitudes as well as healthy eating patterns for life, not faddish diets, are the answer to weight issues.

Do you know the superfoods that will help you beat digestive problems? Discover the foods that have therapeutic power.

In the complex world of food allergy and intolerance, find out the common triggers and how to identify and tackle them.

Foods for your digestive system

The health of your digestive system depends on the inclusion of a number of basic food groups in your diet. Simple nutrients obtained from food can have profound effects on the functioning and efficiency of your GI system.

Your diet is composed of three primary food groups: proteins, fats, and carbohydrates. The first two groups are indirectly essential for the healthy functioning of your GI system—proteins supply the building blocks for the cells, enzymes, and hormones of your digestive system, whereas fats (eaten in moderation) are a vital energy source. They also supply important components for hormones, cell walls, and other parts of your body. The third group of foods, carbohydrates, is used by your digestive system mainly to power the energy-intensive processes of digestion. In addition, one subgroup of carbohydrates—fiber—is particularly significant for your digestive tract, directly affecting its health and function.

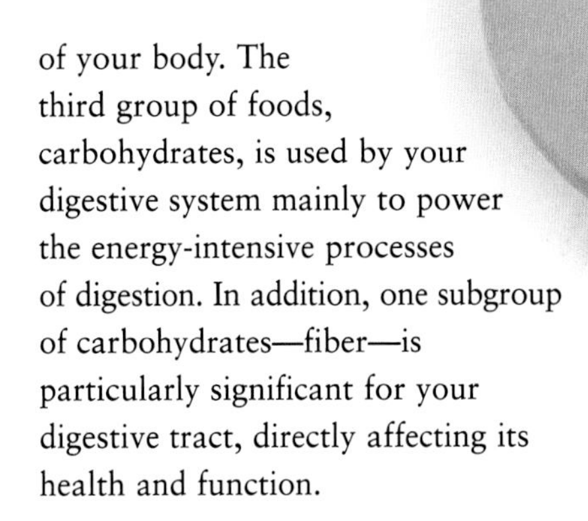

Fresh fruit vs. fruit juice
Fruit juice is a nutritious drink but, it doesn't provide much fiber, most of which resides in the skin of the fruit. An apple provides about 1.6 grams of fiber whereas a 3½-ounce glass of apple juice scores a big zero on the fiber scale.

THE FIBER FACTOR

Doctors first began to find evidence that a high-fiber diet could protect against constipation, hemorrhoids, and colon cancer more than 30 years ago. Drs. Denis Burkitt and Hugh Trowell of Great Britain looked at the diets of people living in rural African villages, who rarely suffered from these problems. The typical diet in these villages was based on unprocessed foods and contained about four times as much fiber as the typical Western diet. The research led doctors to propose that a high-fiber diet (which allows food to pass through the GI system much faster and produces larger, softer feces that are easier to pass) promotes healthy gut function and was responsible for the lower incidence of digestive disease.

What is fiber?

Dietary fiber used to be called roughage. Recently, it's had yet another name change—the correct scientific term now is nonstarch polysaccharide (NSP). Most people, however, continue to use the name fiber. The term describes a number of substances, including cellulose, hemicellulose, pectin, and gum (all of which are found in plant cell walls) plus algal substances (found in seaweeds and algae) and mucilages (from cell secretions). Another plant component, lignin, from the woody part of plants, is a fiber provider.

There are two groups of dietary fiber: insoluble and soluble fiber. Many foods contain both types.

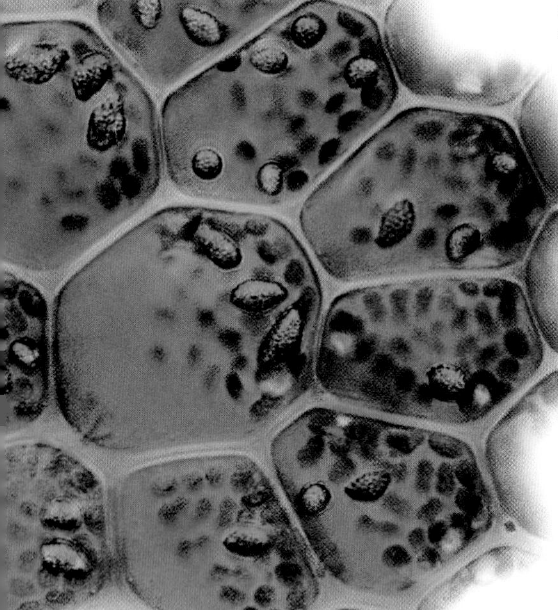

The origins of fiber
Plant cell walls, shown as solid ribs between the polygon cells, are the source of the fiber we get from eating fruit and vegetables. Cell walls contain cellulose, hemicellulose, pectin, and gum, all of which are types of fiber.

Studies carried out at the University of Bristol, in Great Britain, have discovered that butyrate—a fatty acid formed when fiber is fermented by bacteria in the digestive system—can trigger the self-destruction of malignant tumors.

Insoluble fiber

Found mainly in cereals (especially bran and wholegrain wheat), fruit, vegetables, and legumes, insoluble fiber absorbs water and holds it in the large intestine. This accounts for the health-giving properties that Burkitt and Trowell noticed—the resulting stools have a higher water content, making them larger and softer. Insoluble fiber also speeds up the rate at which waste material passes through the digestive system. Faster passage is believed to play an important role in preventing colon cancer by reducing the length of time that cancer-causing toxins are in the digestive system.

Soluble fiber

Great sources of soluble fiber include oats, oat bran, beans, and legumes. This type of fiber can help to lower high blood cholesterol levels by binding to it in the intestines and preventing its absorption. The bacteria that live in your large intestine can use soluble fiber as a substrate for fermentation (their way of making energy); the byproducts are volatile fatty acids, which may be protective against colon cancer, and gas. Soluble fiber also seems to help regulate the absorption of sugar into the bloodstream.

How much fiber do we need?

Most of the adult population in the United States does not eat enough fiber. Although nutritionists recommend an average intake of 20–35 grams of fiber a day, most people manage only 12–17 grams. But reaching the recommended target of 20 grams is easier than you think. The following sample menu for one day provides more than 23 grams of fiber.

- Breakfast of a glass of orange juice (0 grams), muesli with low-fat milk, (3.8 grams) and one slice of whole-grain toast (2.1 grams) adds up to 5.9 grams—a great start to the day.
- A quick and tasty lunch of baked beans (5.1 grams) on whole-grain toast (2.1 grams) clocks up another 7.2 grams.
- An evening meal of chicken casserole, accompanied by a baked potato with skin (4.9 grams) and steamed broccoli (1.7 grams), followed by fresh fruit salad (1.0 grams) adds another 7.6 grams.
- Four dried, ready-to-eat apricots makes a tasty snack whatever the time of day and adds another 2.5 grams.

The fiber "shopping list" to the right rates different fiber providers so you can choose those that are higher in fiber. If you're eating a high-fiber diet, it's best to make sure you're getting your fiber from a wide variety of foods.

Too much of a good thing?

While fiber is important for most people, there are some groups for whom too much fiber can cause problems. People with poor appetites may find it difficult to eat enough because high-fiber foods are very bulky. Some people with irritable bowel syndrome may find a high-fiber diet aggravates this condition.

Young children don't need a high-fiber diet—it will make them feel full without providing enough nutritional value and energy. Too much fiber will cause stomachaches and gas. Children under two need very little fiber; after this age, encourage them to eat fruit and vegetables, which supply fiber in manageable amounts.

FIBER PROVIDERS

PRODUCT	GRAMS OF FIBER PER SERVING
FRUIT	
8 prunes	5.2 g
8 dried apricots	4.9 g
1 medium orange	2.0 g
1 medium apple	1.8 g
1 medium banana	1.1 g
1 tbsp raisins	0.6 g
VEGETABLES	
1 medium baked potato (with skin)	4.9 g
1 portion peas	3.6 g
1 portion carrots	1.4 g
2 tbsp sweetcorn	1.2 g
BEANS AND LEGUMES	
3 tbsp red kidney beans	6.0 g
1 small can baked beans	5.1 g
3 tbsp chickpeas	3.9 g
BREAKFAST CEREALS	
1 bowl (30 g) branflakes	4.5 g
2 Shredded Wheat	4.3 g
1 bowl (50g) muesli	3.8 g
BREAD	
1 slice wholemeal bread	2.1 g
1 slice of whole-grain bread	1.5 g
1 slice white bread	0.6 g
PASTA AND RICE	
Average serving wholemeal pasta	8.1 g
Average serving pasta	2.8 g
Average serving brown rice	1.4 g
Average serving cooked white rice	0.2 g

EASY WAYS TO BOOST YOUR FIBER INTAKE

BREAKFAST Eat wholegrain cereal such as oatmeal, muesli, or branflakes. Choose one that provides 3 or more grams of fiber per serving. Adding fresh or dried fruit provides even more fiber.

SNACKS Fruit, preferably with its skin, is a great way of increasing your fiber intake, and it is an easy snack. Try to eat a minimum of five servings of fruit and vegetables a day.

LUNCH Wholemeal or granary bread with a bowl of vegetable soup is a good high-fiber choice. When buying bread, look on the label for the words wholegrain, wholewheat, or wholemeal. Try to use wholemeal flour for baking wherever possible.

SNACKS Ready-to-eat dried fruits such as apricots or raisins are good snack foods. They are also high in fiber.

DINNER Eat more beans and legumes—try adding them to soups and casseroles. Use brown rice instead of white rice and choose wholemeal pasta.

Bran on everything?

Bran is well known as a rich source of fiber, but sprinkling it over everything you eat to increase your fiber intake is not a good idea. Bran contains high levels of phytate, a substance that binds with minerals such as iron, zinc, copper, and calcium, blocking their absorption. Just two tablespoons of bran a day can be enough to cause a deficiency of these minerals in some people, although this is more likely to be a problem in the elderly and people with poor appetites.

Rather than boosting your fiber intake with bran, it is much better to eat a variety of foods that are naturally rich sources, such as wholegrains, fruit, vegetables, beans, and legumes. This way you'll also be raising your intake of other health-promoting substances—vitamins, minerals, and phytochemicals—at the same time. As another added bonus, people who switch to a high-fiber, low-fat diet often find they lose weight without restricting their food intake. High-fiber foods are bulky and slow down the rate at which the stomach empties, helping you to feel full for longer.

Too much too soon

If you're not accustomed to eating a high-fiber diet, suddenly increasing the amount you eat can cause bloating and gas. Your digestive system needs time to adjust, so increase your fiber intake gradually and give your body time to adapt. If you do this, painful gas and bloating won't be a problem. If your current diet is not high in fiber, it can take a few weeks for the enzymes needed to digest fibrous foods to develop; if undigested foods pass into the large intestine, they are fermented by the gut bacteria, producing gas. Gas-generating food culprits include beans and legumes, cabbage, brussels sprouts, grapes, raisins, and apples.

The maximum capacity of the colon is approximately 8¾ to 10½ pints of water; on an average day the colon absorbs between 1¾ and 3½ pints.

FIBER AND WATER—A HEALTHY PARTNERSHIP

To do its job properly, fiber has to have fluid. Insoluble fiber passes through the digestive system pretty much unchanged. Once it arrives in the large intestine, however, it acts like a sponge, absorbing large quantities of water—up to 15 times its own weight. This process helps to prevent constipation and problems such as diverticular disease by creating bulky, soft stools that are then easy to pass.

Food combining—diet fact or fiction?

TALKING POINT

Food combining—or the Hay diet as it is sometimes known, after its inventor Dr. Howard Hay—is based on the idea that proteins and carbohydrates should not be eaten together at the same meal. The theory is that these food groups require different conditions for digestion, so if you eat them at the same meal, inefficient digestion results, leading to a build-up of toxins in the body. Many people claim that abiding by the Hay diet gives them increased energy and a sense of well-being, although there is no scientific evidence to support this.

Your large intestine absorbs water from drinks and food. In a low-fiber diet, more water can be absorbed from food as it moves slowly through the GI system. Upping your fiber intake (especially insoluble fiber) means that the fiber "competes" with the large intestine for available water, so less is absorbed. Aim for a fluid intake of at least 3 to 3½ pints every day to satisfy your body's need for extra water when on a high-fiber diet.

CARBOHYDRATES FOR ENERGY

Carbohydrate is a term used to describe a type of food composed of sugar molecules. Carbohydrates include simple sugars, such as glucose and sucrose, and complex substances such as starch (known as complex carbohydrates). Foods rich in carbohydrates also supply fiber, protein, vitamins, and minerals, but their main job is to provide energy.

For a healthy diet, experts recommend that at least 40 percent of your daily energy requirements should come from high-starch carbohydrate foods such as bread, rice, breakfast cereals, pasta, and potatoes. Choose fiber-rich varieties such as wholemeal bread and wholegrain cereals whenever possible—these help to keep the digestive tract working efficiently and provide a slow-release form of energy that helps to keep blood sugar levels stable.

Starchy foods are not fattening

Some people mistakenly believe that starchy foods are fattening. In fact, they are low in fat and are also good sources of fiber. It is only when they are eaten in conjunction with fat—pasta in a rich creamy sauce, potatoes fried in oil—that these foods become high in calories. A 4-ounce serving of boiled potatoes contains only 80 kilocalories, but the same weight of potato chips contains 250 Calories!

ARE VEGETARIAN DIETS HEALTHIER?

Vegetarianism is becoming increasingly popular. One study shows that 2.5 percent of the U.S. population is vegetarian. Studies have shown vegetarians are less likely to suffer from several diseases including diverticular disease, irritable bowel syndrome, hemorrhoids, and colorectal cancer. As with any diet, there are healthy and balanced, as well as unhealthy vegetarian diets. But on the whole, vegetarians tend to have diets that are higher in complex carbohydrates and fiber than meat eaters, which probably explains why they are less likely to suffer from intestinal problems.

BALANCING FOODS IS KEY

Remember that although fiber and carbohydrates are crucial for a healthy digestive system, eating a balanced diet containing protein, vitamins, and some fat is vital for your body to function properly. Don't go to extremes and fill up completely on fiber-rich food.

A healthy GI system for life

Maintaining a healthy GI system doesn't depend simply on what you eat, but it also depends on how you eat. Good eating habits benefit the long-term well-being of your digestive system, and the sooner you start, the better off you'll be.

The process of eating is something that most of us take for granted, so it may come as a surprise to learn that the way you eat—from the timing of your meals to your activities while you are eating them—can affect your digestion and, in turn, your general health. Developing good eating habits is a major step toward a healthy lifestyle, benefiting your digestive system both directly and indirectly. Habits learned during childhood will last a lifetime.

Babies have more taste buds than adults, not just on the tongue but everywhere in the mouth. They also prefer bland foods, whereas adults will tend to enjoy many flavors as they "acquire" tastes.

THE FIRST FOOD

It's great to know that the first food we have is perfectly formulated for an immature digestive system. Breast milk and formula milk both contain all the essential nutrients and are easily digested to release what a tiny body needs. Healthcare providers and doctors always encourage mothers to breastfeed for at least a few weeks because it helps to boost the newborn's immune system and stimulates bonding. But even if breastfeeding is not a success, a baby will thrive on formula milk.

STARTING ON SOLIDS

Weaning, sometimes called "mixed feeding" or "starting on solids," is the gradual process that begins when parents start to introduce solid foods into their baby's diet. Throughout weaning, milk should continue to be an important food, but as a baby's digestive system gears up, it is vital to start to gradually increase the amount, texture, and variety of foods in his or her diet.

The process of weaning takes place over many months. At first (from the age of four to six months), foods are introduced one at a time to avoid upsetting the immature digestive system. After about six months you should be able to combine foods for your baby—although it's still best to introduce new foods one at a time before mixing them.

By about 9 to 12 months, as the digestive system matures and becomes more able to cope, a wider variety of foods of different types and texture can be added. Around the time of a baby's first birthday, the digestive system should have developed enough to be able to handle much the same kinds of foods the rest of the family eats. Remember, however, not to include any salt in your baby's food.

Why do babies need weaning?

As babies get older, milk alone cannot satisfy all their nutritional needs. Their stores of certain nutrients—such as iron, which is built up from the mother's supplies

A LIFETIME OF EATING

The food we eat changes throughout our lives, from our first feedings to the adventures of weaning and the discovery of enticing new foods, which can continue well into adulthood and later life.

It's not just what we eat that changes, but also how we eat and how often. This is dictated in part by the way our digestive systems handle food. For example, the large meals that we may easily have coped with in our young adult lives can become a cause of indigestion; consequently older people often opt for smaller, more regular meals. Social and cultural influences also play a large part in what and how we eat—as we enjoy making mealtimes a special occasion with our friends and families.

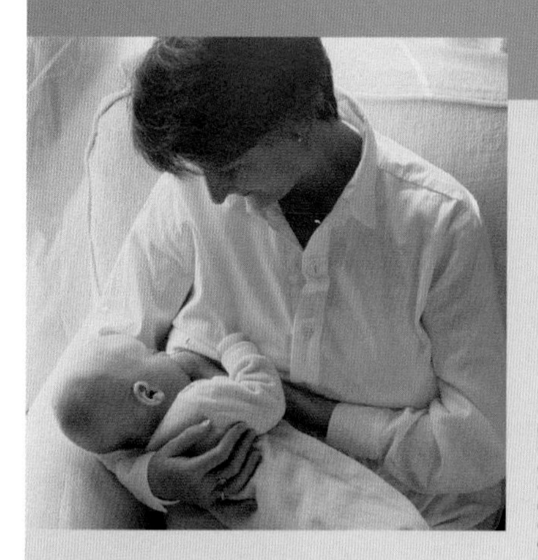

Milk—the first food of life
Our first eating habits are very much a one-on-one experience with our parents, usually our mothers, at either the breast or the bottle.

Finger foods

Once toddlers have teeth, chewing foods is no longer a problem. Finger foods allow toddlers to enjoy feeding themselves, and older children can enjoy these as snacks—preferable to cookies and salty snacks.

Developing exotic tastes

Teenagers may start to become more adventurous in their eating habits. A child who may have shunned any type of spicy food may become a teenager who tops everything with hot chili peppers.

Anyone for cake?

As we grow older, we tend to eat little and often, rather than three large meals throughout the day. Some people also develop a definite "sweet tooth" as their taste buds become less sensitive.

of iron before birth—start to run out and need to be supplemented by a more varied diet. Energy needs also increase. In addition, moving on to more solid foods also helps to develop the muscles necessary for chewing and for speech.

The process of weaning usually begins some time between the ages of four and six months, but the exact age depends entirely on the individual baby. Every baby is different and what suits one baby isn't necessarily right for another.

TODDLER TAMING

As children get older, it's not unusual for them to become picky about what they eat. They soon learn that refusing to eat is a powerful weapon and a good way of gaining attention. Usually, pickiness is only a passing phase, but most parents are understandably concerned about the immediate effects of food refusal on their child's general health. This period is also significant in the longer term because it can be a time when deep-seated attitudes toward food and eating are laid down; these include associations between food refusal and control, or between eating and attention seeking.

Be food wise

Don't allow mealtimes to turn into a battle of wills. Never bribe, force, or coerce a child to eat—this can make the problem worse. If your child refuses a certain food or meal, simply clear it away without any comment. Be aware that children often pick up bad eating habits from other members of their family. Discourage older siblings from being fussy and make sure that your own eating habits set a good example.

IDEAS FOR A HEALTHY PACKED LUNCH

HAM AND PEA TORTILLA

Make a tortilla using eggs, ham, and peas; complement the tortilla with a nutritious and fiber-rich carrot salad with dressing. Add some fruit juice for a complete and healthy meal.

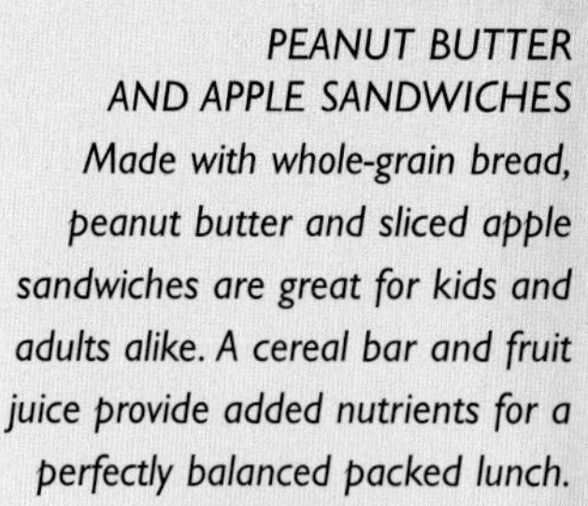

PEANUT BUTTER AND APPLE SANDWICHES

Made with whole-grain bread, peanut butter and sliced apple sandwiches are great for kids and adults alike. A cereal bar and fruit juice provide added nutrients for a perfectly balanced packed lunch.

PITA BREAD LUNCH

Pack a whole-grain pita bread with hummus and coleslaw and add a package of low-fat chips, a small box of raisins, a kiwi fruit, and juice or sugar-free drink—perfect for kids.

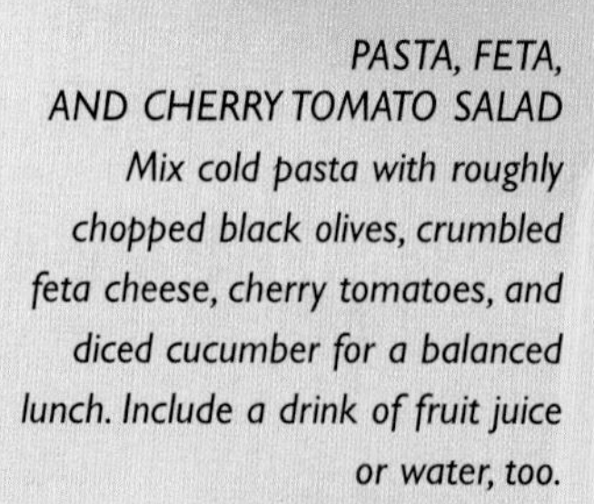

PASTA, FETA, AND CHERRY TOMATO SALAD

Mix cold pasta with roughly chopped black olives, crumbled feta cheese, cherry tomatoes, and diced cucumber for a balanced lunch. Include a drink of fruit juice or water, too.

Snacks aren't all bad—young children may need "pick-me-ups" between meals for sufficient energy. Offer sandwiches, fruit, scones, or a fruit yogurt rather than cookies or salty snacks.

HEALTHY EATING FOR OLDER KIDS

Older children and teenagers often eat a lot of unhealthy snacks and processed foods. You can help to improve their diets by providing them with nutritious packed lunches, but preparing healthy food that children and teenagers will enjoy can be a challenge.

The main rule is to make food quick and easy to eat. Don't fall into the routine of providing the same packed lunch day after day—try to make the meals fun and imaginative. With children, it's important that food looks attractive—raw vegetables washed and cut into strips are colorful easy to prepare and eat, and they provide all-important vitamins and minerals.

Sandwiches are the mainstay of most packed lunches. It's best to use whole-grain or granary bread, but if your child doesn't like them, use high-fiber white, or compromise and use one slice of white and one slice of wholemeal. Introduce variety by using a selection of bread and rolls. Lightly toasted pita bread splits easily to provide a natural pocket to take generous portions of salad and a range of different fillings.

Most children want something sweet in their lunch boxes. Home-made cakes and cookies made with whole-grain flour are a healthier option than store-bought varieties. Dried fruits such as raisins and apricots are also good choices.

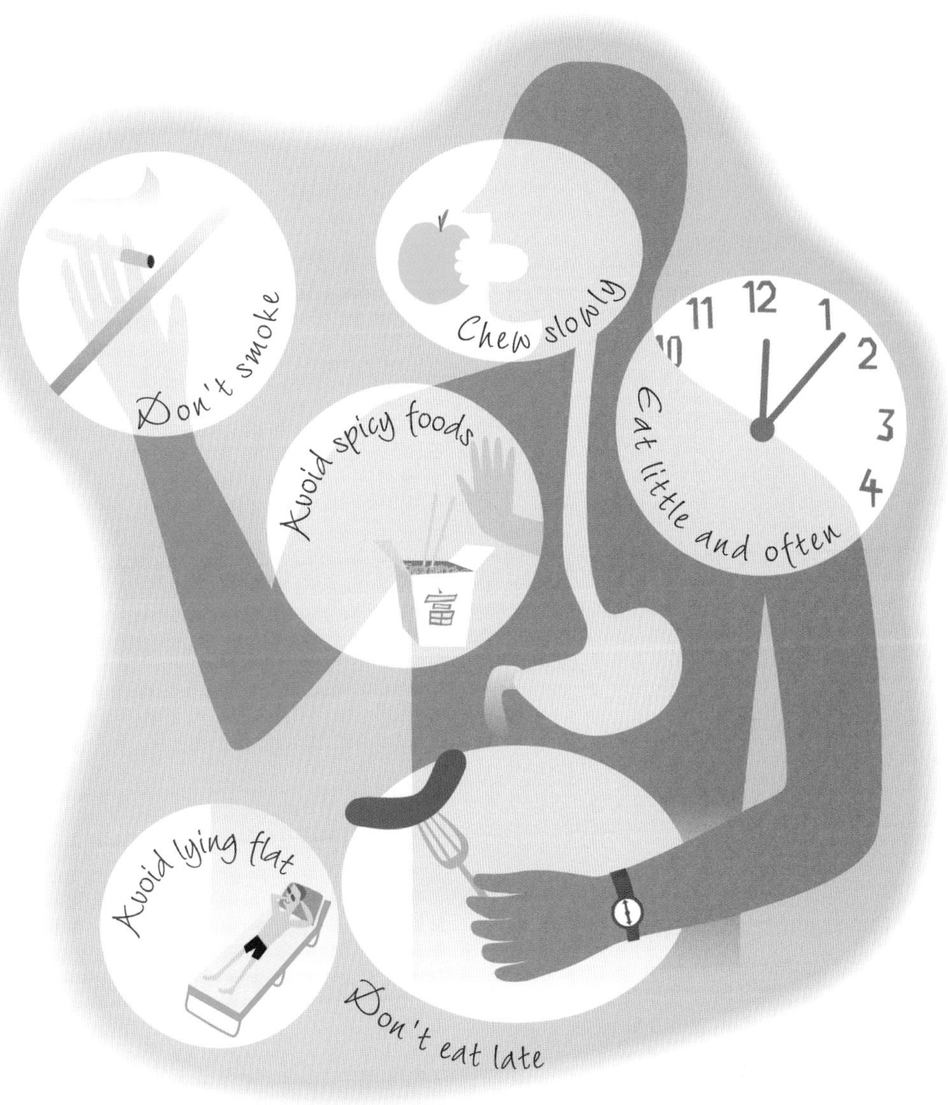

Avoiding indigestion and heartburn
Try some of these tips to reduce the chance of having indigestion and heartburn. As well as thinking about what you eat, also consider how and when you eat—eating too quickly or late at night are common causes.

DEVELOP HEALTHY ATTITUDES TOWARD EATING

Eating habits developed in childhood lay the foundation for your attitude toward food as an adult. But, it is never too late to change your eating habits.

As well as paying attention to the kinds of food you eat, there are a number of tips you can follow to improve the way you eat. These simple measures can help control your appetite and calorie intake; they can also help make your diet more balanced and give your digestive system a helping hand.

Focus on your food

If you are distracted when you eat—by watching television, for example—you are are more likely to miss your body's "fullness" signals. It takes about 15 minutes for messages from your stomach to turn off your brain's eating drive. If you eat too quickly, your stomach fills up before your brain knows that you are full, and you will overeat, ending up feeling uncomfortable.

Try to make the time to sit down and eat your meals at a table—even

FOR THE OVER-80s

How to counteract poor appetite

Many older people find they don't feel like eating as much as they did when they were younger. Poor appetite can be a problem if it leads to inadequate nutrition.

- **Have a drink** A small glass of sherry or your favorite alcoholic drink half an hour before you eat might help to get your appetite going. If you are taking any medicines, check first with your doctor or pharmacist.
- **Exercise** A short walk before your meals will help to stimulate your appetite.
- **Eat smaller meals** Three small meals a day will provide good nutrition in more manageable amounts than one or two larger meals will.
- **Pop a vitamin pill** As you age, it's probably a good idea to take a multivitamin and mineral supplement. Ask your doctor or pharmacist for a recommendation.
- **Use "supplemental" drinks** In addition to being nutritious snacks, the unflavored varieties can be added to soups to enrich them.

if you eat alone—rather than grabbing food while you are busy doing something else. If you eat on the move, you are more likely to get indigestion and heartburn because blood that is needed for digestion is diverted away from the digestive system toward the muscles.

Chew properly

Chewing food thoroughly gets the whole digestive process off to a good start. It breaks down food into smaller, more manageable pieces, and mixes food with saliva, beginning the breakdown of carbohydrates into simple sugars. Chew each mouthful carefully before you begin the next. If you eat too quickly, you swallow air, and this can lead to painful gas.

Less tea with meals

Caffeine and tannin—a substance found in abundance in tea—interfere with the absorption of iron from food as it passes through the GI system. Iron deficiency can result in tiredness, lack of energy, and increased susceptibility to infection.

Cutting out tea, coffee, and cola drinks with meals could help to boost your iron reserves. Vitamin C increases the absorption of iron from foods, so eating fruit and vegetables or drinking fruit juice at the same time as iron-rich foods (such as lean red meat or spinach) is a good idea.

Delay brushing

Don't brush your teeth immediately after eating or drinking acidic foods. The acid from fruit juice and foods like pickles can weaken the enamel of your teeth. Brushing your teeth right after consuming such foods or drinks can make things worse. Instead, rinse your mouth with a little water and wait at least 45 minutes before brushing.

CHANGES IN APPETITE AND DIETARY NEEDS

As we grow older, physiological and psychological changes occur that have a direct effect on our nutritional needs. The body becomes less efficient at absorbing and using many vitamins and minerals. In addition, the long-term use of some prescription drugs can reduce the ability of your body to absorb certain nutrients. At the same time, many people find that as they get older, their appetites decrease. Yet the need for vitamins and minerals stays the same with age, or in some cases increases. So, it becomes even more important for you to make sure that the food you do eat is healthy and that it supplies all the important nutrients your body needs.

Spice up your life

As you get older, your senses of smell and taste becomes less acute, and this can have a negative effect on both your appetite and your enjoy-

Good guys
The Lactobacillus acidophilus *bacterium is one of the "good" bacteria that thrive in the healthy GI system. They help to keep harmful bacteria in check and contribute to efficient digestion.*

ment of food. Try compensating for these effects by enhancing the flavor of your meals with herbs and spices (don't overdo the spices), garlic, lemon juice, mustard, or vinegar.

Enjoying your food

Eating should be a pleasure, so it's worth making an effort each day to plan your meals and enjoy them. Choosing a healthy diet doesn't have to be expensive or involve elaborate cooking. A well-planned snack can be just as nutritious as a cooked meal, and frozen vegetables and canned beans are as nutritious as fresh ones. Different foods provide different nutrients, so eating a varied diet is the best way to ensure that you get a good mix of all the necessary nutrients.

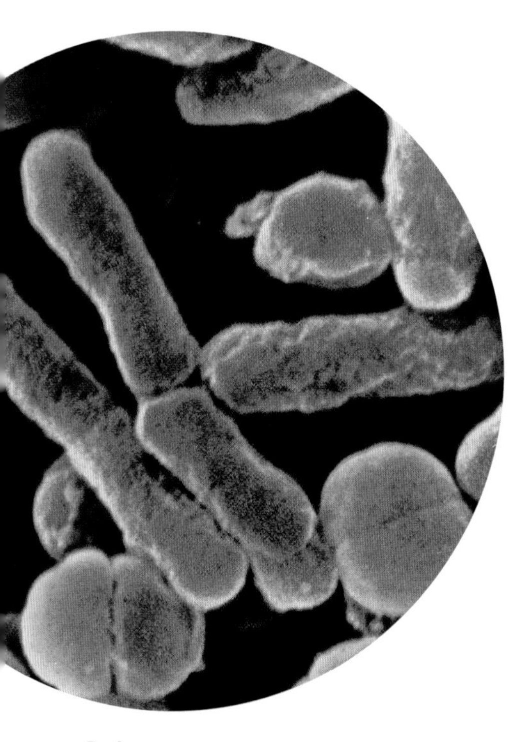

Bad guys

Although prolific in your intestines, particular strains of the Escherichia coli *bacterium (the rod-shaped organisms) and streptococci (the spheres) can be potentially harmful and cause gastrointestinal infections.*

CULTIVATE HEALTHY GI FLORA

At birth, a baby's sterile digestive tract picks up bacteria from its mother, and within a few hours the newborn's intestine is home to millions of friendly bacteria collectively known as the GI flora. By the time you reach adulthood, your GI flora number in the trillions, with more than 200 species of bacteria competing for the resources that you provide by eating.

A balanced GI flora

Your GI flora include "good" bacteria, the most prolific of which is a harmless strain of *Escherichia coli,* which helps in the digestion process and produces important nutrients, and "bad" bacteria, which can cause illness. These good and bad bacteria coexist in a fine balance, which is extremely vulnerable to disruption. Stress, poor diet, or taking antibiotics can all upset the bacterial balance, allowing the "bad guys" to gain the upper hand. Problems such as candidiasis (see below) may result.

Bacterial imbalances

Candida albicans is a yeastlike organism that occurs naturally in the mouth, digestive tract, and vagina and on the skin. Normally, it causes no problems because it is kept in check by "friendly" bacteria. If the immune system is weakened, however, or if the protective bacteria in the gut are destroyed by stress, poor diet, or antibiotics, *Candida albicans* can grow unchecked, causing a condition known as thrush (usually in the mouth or vagina).

Alternative health practitioners relate an overgrowth of *Candida* to bloating, diarrhea, and constipation.

Probiotic foods

You can take certain steps to keep the balance in favor of the good bacteria and give your GI flora a helping hand. One such step is to eat what are known as probiotics—foods or supplements that contain "friendly" bacteria. Eating probiotics regularly delivers a fresh supply of good bacteria, preventing the growth of the bad guys that can cause problems. Probiotic foods include live yogurt, which supplies "friendly" *bifidus* or *acidophilus* bacteria, and fermented foods such as sauerkraut, buttermilk, and miso.

Prebiotics

Eating foods that promote good bacteria is another way of addressing the balance of bacteria in your GI flora: They stimulate the growth of good bacteria in the GI system while inhibiting the growth of bad bacteria. Prebiotics include the dietary fiber oligosaccharide, found naturally in Jerusalem artichokes, onions and chicory. Food manufacturers are looking into the possibility of adding these ingredients to a range of foods.

Some also believe that in the digestive tract, *Candida albicans* can cause damage to the mucus that lines the GI tract, allowing waste products to leak into the blood, a condition that has been called GI dysbiosis.

Complementary practitioners believe that candidiasis and dysbiosis are extremely common and cause a range of problems such as chronic fatigue, headaches, skin problems, and depression. However, this is not based on any scientific evidence or on properly controlled trials.

Banish harmful microorganisms

Candida albicans feeds on sugars and refined carbohydrates in the digestive tract. For this reason, one way of treating candidiasis is to follow a strict sugar-free diet. At the same time, eating prebiotic foods and probiotic supplements can encourage the GI system to become recolonized with friendly bacteria. Other foods helpful in combatting candidiasis include garlic and dark green leafy vegetables—these all have natural antifungal properties. Olive oil, which is rich in oleic acid, is also thought to be helpful because it inhibits the growth of *Candida albicans*.

Should you eat the foods you crave?

Some people believe that cravings are the body's way of making sure that you eat foods providing specific nutrients that are lacking in the diet. There is, however, very little scientific evidence to support this view. Whether or not you should succumb to cravings depends to some extent on how often they occur. Provided your diet is generally healthy and balanced, go with the flow. If your cravings become bizarre or you're lusting after certain foods all the time, discuss this with your doctor.

ASK THE EXPERT

FOOD CRAVINGS

Many women find that their monthly menstrual cycles are linked to food cravings. One in four women notices an increase in appetite just before her period is due—this is often accompanied by a craving for carbohydrates, in the form of sweets and, in particular, chocolate.

Chocoholics

Some scientists suggest that a craving for chocolate is caused by desire for a compound called phenylethylamine, a substance that stimulates the release of the "feel-good" hormone dopamine. There is some evidence that the minerals chromium and magnesium can help ease this sort of craving. Try adding foods to your diet that are naturally rich in these minerals. Foods to include are wholegrain cereals, brewer's yeast, meat, kidneys, cheese, nuts, legumes, bananas, and green leafy vegetables.

The food of cravings
Chocolate is the focus of many food cravings and is believed to stimulate release from the brain of "feel-good" hormones.

Cravings during pregnancy

Women often experience weird and wonderful food cravings during pregnancy, probably an effect of changing hormone levels. Most of these food cravings are nothing to worry about. Be aware of your general diet, however, and make sure you eat all the nutrients you need.

Craving for nonfood substances (such as clay or paint) is known as pica and can be a sign of iron deficiency. You should mention it to your doctor or midwife. Make sure your diet includes enough iron-rich foods and vitamin C, which aids iron absorption.

Theories on cravings

Some people believe that cravings or aversions are signs of food sensitivity (see pages 78 to 82). For example, a person who hates milk may be lactose intolerant; others find they crave the substance that causes the sensitivity. Another theory is that your body craves foods that contain nutrients in which it is lacking, but no study has proven this to be true.

A healthy attitude toward weight

Maintaining a consistent, healthy weight benefits your general well-being and the health of your digestive system. Diet fads aren't the answer to weight problems; the key is to develop a sensible attitude toward weight and eating.

Weight is a modern obsession. In theory, this obsession could encourage people to maintain a healthy weight. But, in practice, the incidence of obesity has grown to epidemic proportions and more and more people have health-damaging attitudes toward food and eating.

The issue of weight is of particular importance for your digestive system, because its workings and your body weight are intimately related. Being overweight can seriously affect your GI health, while looking after your digestive system is one of the best ways to control your body weight.

WHAT IS A HEALTHY WEIGHT?

Our views on what constitutes an ideal weight and shape have changed dramatically over recent decades. In the 1950s Marilyn Monroe was considered to have the perfect female form. But today's waif-like catwalk models are a far cry from her more generous size-14 figure.

Magazines and fashion designers may promote thin models, but the true ideal weight is that at which people remain healthiest and live longest. Check out whether you are a healthy weight using the chart below.

Weight expectations

Although no one can deny that obesity and being overweight are growing problems in many Western countries, many women—and a growing number of men—have unrealistic ideas about what constitutes an ideal weight. Obsession with weight can itself cause serious health problems. A CDC study of students in grades 9 to 12 found that girls (58.5 percent) were less likely than boys (68.8 percent) to consider themselves the correct weight.

How you feel about your weight can have a fundamental effect on your self-esteem and happiness and can have an important, long-lasting impact on your health. Eating disorders such as anorexia nervosa and bulimia can result in severe nutritional deficiencies and long-term problems like osteoporosis.

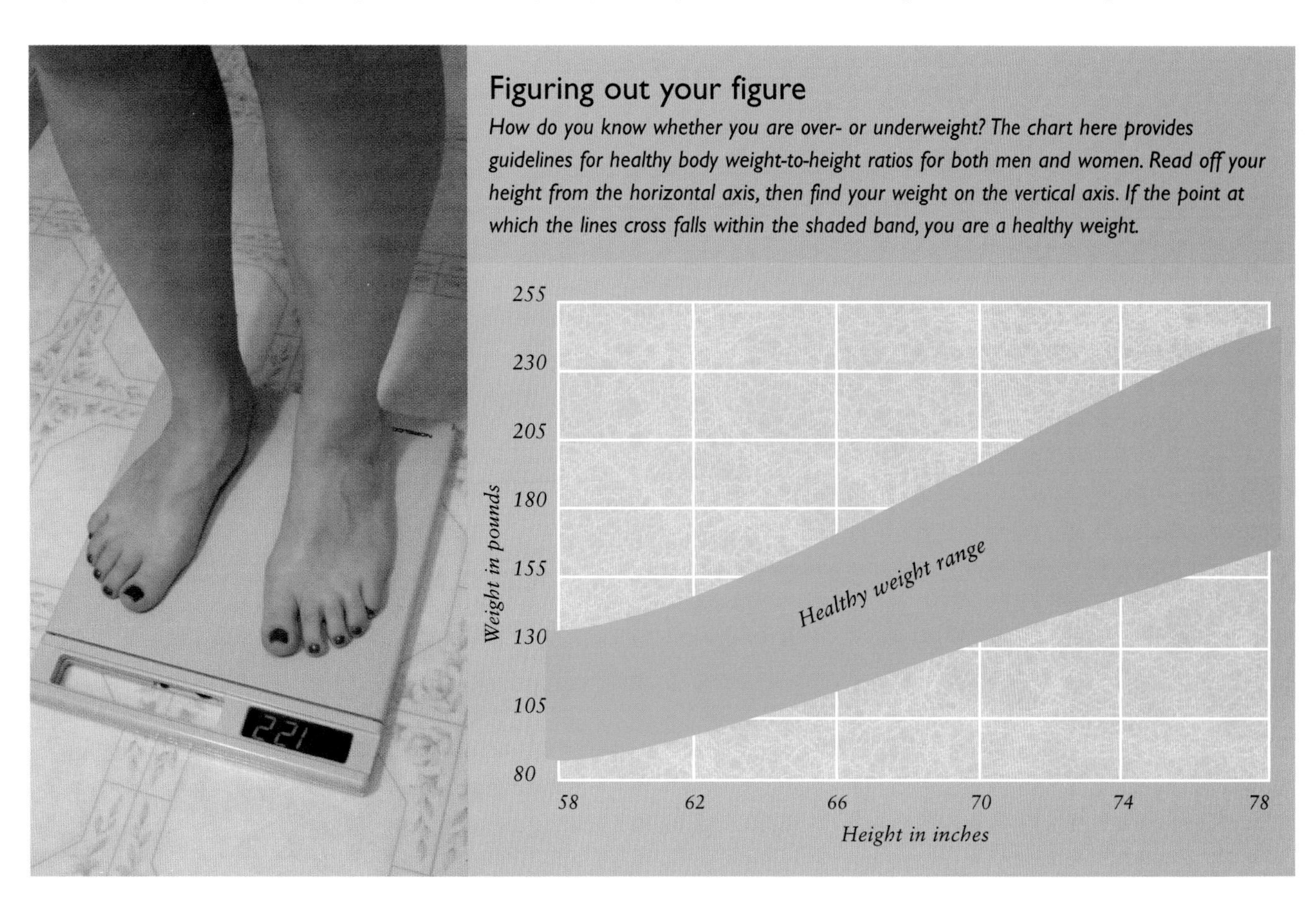

Figuring out your figure

How do you know whether you are over- or underweight? The chart here provides guidelines for healthy body weight-to-height ratios for both men and women. Read off your height from the horizontal axis, then find your weight on the vertical axis. If the point at which the lines cross falls within the shaded band, you are a healthy weight.

MAKE SMALL CHANGES FOR A MAXIMUM EFFECT ON YOUR WEIGHT

Making small changes in the way you eat and the type of food you choose can have a significant impact on your weight. The good news is that this doesn't have to mean missing out on your favorite foods. In fact, it is important to include foods that you enjoy eating. A diet that leaves you feeling deprived, unhappy, or dissatisfied is a diet that's very quickly going to be abandoned.

Reduce the amount of high-fat food you eat. Try to combine any high-fat foods with high-fiber foods—fiber helps to reduce the absorption of fat by your intestines. For example, eat whole-milk cheese with whole-grain bread or yogurt with slices of banana.

Never skip meals and don't allow yourself to get too hungry. If you do, you'll be more tempted to snack and more likely to overeat at your next meal. Aim to eat three small- to medium-sized meals. Using smaller plates can help you stick to smaller portions!

Be prepared for snacking urges—make sure your cupboard is full of healthy foods, and have plenty of low-calorie snacks, such as fruit, available.

Choose low-fat cooking techniques as much as possible—poach, braise, steam, dry-roast, grill, or stir-fry.

Freeze or throw away leftovers as soon as you've finished eating. That way you won't be tempted to come back for more.

Build some exercise into your daily routine. For example, walk up the stairs rather than taking the escalator or elevator. If you travel by bus, hop off one stop earlier than usual and walk the rest of the way.

Try to reduce the amount of meat you eat and choose lean meat over fatty types. Trim away any visible fat from meat before cooking. When preparing or eating poultry, remove the skin first—this is where most of the fat lies.

Make a list of food you want to buy, and try to stick to it when you go shopping. Don't be tempted to buy chips and cookies if you know you won't be able to resist them the minute you get back home.

Can detox diets help in losing weight?

People use detox programs for a number of reasons, which may include weight loss. These types of diets are said to rid the body of toxins and poisons and allow the digestive system a rest. Most doctors, however, claim that detox diets are unnecessary and potentially dangerous. Any weight loss that is achieved is likely to be temporary (it's just water) unless you adopt a new set of healthy-eating habits for life.

ASK THE EXPERT

Health risks of being under- or overweight

There's no doubt that being overweight increases the risk of several serious health problems. Heart disease, diabetes, high blood pressure, stroke, reproductive disorders, gallstones, osteoarthritis, and certain types of cancer are all more common in people who are overweight. People who are underweight also have increased health risks, such as a greater risk of infection and poor general health.

In particular, being overweight can affect your digestive system. Hiatal hernias, heartburn, and indigestion are all more common in people who are carrying too much weight. Several simple measures can help you avoid heartburn and indigestion (see page 69). But, for many people, the most effective solutions to these common digestive problems is to lose the excess weight.

SLOW BUT SURE

The best and safest way to lose weight is slowly and steadily—between 1 and 2 pounds a week is the ideal rate. If you lose too much weight too quickly, there's a danger that you may lose lean muscle tissue as well as fat. Your metabolic rate is related to the amount of lean muscle tissue you have, which means that losing muscle lowers the rate at which you burn calories, making it even harder to lose weight.

To achieve steady, healthy weight loss, you need to reduce your Calorie intake to around 600 Calories less than the average recommended daily intake. Between 1,200 and 1,500 Calories a day should be sufficient for most women; men need 1,750 to 1,950 Calories a day. It's important that you spread your Calorie allowance throughout the day—eating little and often helps to maintain blood sugar levels and makes it easier to resist the urge to snack or binge.

Long-term change equals long-term success

Losing weight is relatively easy—keeping that weight off is the real challenge. Most diets provide a short-term fix to what is a long-term problem, with the result that many people who lose weight quickly regain it once they stop dieting.

If you really want to lose weight, you need to change your eating habits on a long-term basis—crash diets don't teach you how to change these. So forget about dieting—instead, think about a whole new way of eating. Successful dieters are those people who learn how to change both their eating habits and their attitude toward food on a permanent basis.

TRYING TO GAIN WEIGHT

Gaining weight can be as difficult as losing it. The healthiest route to weight gain involves a combination of muscle-building exercise and a balanced diet. Choose high-protein foods (such as lean meat, fish, and dairy products), fiber-rich carbohydrate foods (such as whole-grain bread and pasta) and foods rich in "unsaturated" fats (such as oily fish, nuts, and seeds).

Eating little and often will help to increase the total quantity of food you can manage. Try "supplemental" drinks or smoothies between meals.

Blueberry and watermelon smoothie
Nutritious smoothies—fruit, milk, and yogurt—can boost your Calorie intake as "mini meals" between meals; the banana variety is especially satisfying.

Eat to beat digestive problems

From the homely oat to the exotic fennel seed, nature has provided a medicinal treasure trove with the therapeutic power to help solve many digestive ailments and maintain the good health of your digestive system.

Drink tea
In animal studies, tea has been shown to help protect against certain types of cancer. Evidence that it has a similar effect in humans looks promising.

Around the late fifth century B.C., Hippocrates, the father of Western medicine, advised his readers, "Let food be your medicine, and medicine be your food." The ancient Greek sage was building on a long tradition of using food as a potent therapeutic tool, and of using the digestive system as a vital delivery route for that therapy.

FOOD SOLUTIONS

Hippocrates was particularly keen on using food to treat digestive problems. He advocated wheat and barley bran to cure constipation, for instance, a strategy that still makes good sense today. Simply by eating and digesting food, you can deliver therapeutic foods directly to the site of trouble.

Nowadays a lot of food medicine is known to us as folk wisdom—you've probably had chicken soup fed to you to treat a cold or heard of stewed prunes as a food that promotes regular bowel movement. But there are plenty of other foods of which you may not be aware that have health-giving properties that benefit your digestive system.

DIGESTIVE *SUPERFOODS*

Certain foods are great for keeping your digestive system in good working order. By including some of these superfoods in your day-to-day diet, your GI system will reap the benefits.

ASPARAGUS This vegetable is rich in fructoligosaccharides, which stimulate the growth of "friendly" gut bacteria. In turn, these bacteria aid digestion.

PRUNES A substance called hydroxyphenylisatin, found in prunes, stimulates muscles in the bowel. This substance accounts for the mild laxative effect of this dried fruit.

SPINACH Several leafy green vegetables, including spinach, contain nitrates. These chemicals are converted within the body to a potent antibacterial chemical that helps to protect the digestive tract against infection.

TOMATOES These are another source of fructoligosaccharides and encourage the growth of "friendly" GI bacteria.

OATBRAN As well as being a good source of several important vitamins and minerals, oatbran contains fiber, which speeds the passage of waste products through the large intestine.

Foods for general digestive health

Probiotic and prebiotic foods, such as yogurt, onions, and Jerusalem artichokes, help to maintain healthy GI flora and fight problems like thrush and candidiasis (see page 71). There is even evidence that yogurt can strengthen the immune system.

Studies that look at the incidence of colon cancer within different population groups suggest that people who eat more fruit and vegetables are less likely to get this type of cancer. Several studies show that a diet rich in fiber can help prevent problems such as constipation, diverticular disease (a condition in which part of the lining of the colon forms small inflamed pouches), and hemorrhoids, as well as some types of cancer.

EATING TO BEAT CANCER

It has been estimated that up to one third of all cancers in Western countries are associated with poor diet, and cancers of the digestive system are among the top killers. Experts suggest that dietary changes could prevent nine out of ten deaths from colon cancer, for instance. Fiber speeds the passage of food and may prevent potentially cancer-causing substances from remaining long enough to do harm.

Fruit and fiber

Fruit and vegetables are rich in antioxidants—substances that mop up harmful chemicals called free radicals and help your immune system to prevent and fight cancer. Fruit and vegetables are particularly effective at guarding against several digestive cancers, including that of the stomach, large intestine, and esophagus.

Dietary fiber (see pages 62 to 65), from sources including fruit, is also important in the prevention of colon cancer. Official guidelines indicate that we should eat at least five servings of fruit and vegetables a day to maintain good digestive health.

A soy solution?

There is a growing amount of evidence to suggest that soy may play a major role in preventing certain types of cancers, including colon cancer. In countries where soy is an important source of protein in the diet, the population has a reduced incidence of colon cancer.

When you choose soy you may want to consider whether it has been genetically modified (GM). Soy was one of the first crops to be altered genetically, but in the late 1990s claims were made that GM foods may be harmful to health. Until further research has cleared GM foods, you may want to find a supply of soy that has not been subject to modification.

Food remedies for digestive ailments

Peppermint and mint tea Infusing fresh mint or peppermint leaves in boiling water for a few minutes makes a refreshing drink that is said to improve digestion and relieve indigestion.

Fennel and fennel seeds Chewing or eating fennel or fennel seeds may relieve indigestion and stomach cramps. Fennel tea may relieve nausea, flatulence, and bloating. Avoid these if you are pregnant because they may encourage menstruation.

Water This remains one of the best solutions for constipation or bad breath (which can be caused by constipation). You should aim to drink at least 8 to 10 glasses of water a day.

Live yogurt Eating live yogurt encourages "friendly" bacteria to flourish in the GI and genital tracst. It can help to control thrush, caused by an excess growth of the yeast *Candida albicans*.

Ginger Whether sucked as a piece of the crystallized form or drunk as ginger ale or tea, ginger quells nausea. To make ginger tea, grate a ½-inch piece of root ginger into a mug of boiling water and leave to stand for 10 minutes.

Manuka honey Taken after meals and at bedtime, this honey—from the New Zealand tea tree— is said to help heal stomach ulcers and kill the *Helicobacter pylori* bacteria that cause them.

Cardamom This spice is said to be good for digestive problems such as stomach cramps, indigestion, and flatulence. It helps prevent acid reflux and belching.

Food allergies and intolerance

In recent years there has been a growing awareness that eliminating certain foods from the diet may be the key to successful treatment of a range of health problems linked to food allergies and food intolerance.

Allergies are an increasingly common health issue in the developed world. Conditions such as hay fever and asthma are among some of the most common childhood ailments, and these are only the most well-known types of allergies. According to certain schools of thought, allergic, or allergic-type, reactions to foodstuffs, additives in foods, and chemicals in the environment are the underlying cause of a number of chronic (meaning long-term or constantly recurring) conditions ranging from arthritis to depression.

A COMPLEX ISSUE

Unfortunately, this area of medicine is highly contentious. Conventional medicine has a narrow definition of allergy and only recognizes a type of food reaction known as immediate food allergy, which has clear-cut symptoms and is comparatively rare. Alternative allergy practitioners believe that a closely related type of reaction, known as hidden food allergy, is responsible for a wide range of vague symptoms and is one of the most common causes of illness in the Western world.

To complicate the picture further, a type of food reaction called intolerance, or sensitivity, also exists. Such reactions have a different mechanism to allergies, but the terms are often used interchangeably.

ALLERGY OR INTOLERANCE?

Although food allergies and food intolerance can produce very similar symptoms, they occur as the result of different mechanisms.

The allergic mechanism

Both conventional and alternative allergy practitioners agree on the basic mechanism involved in food allergy: A food allergy occurs when the immune system overreacts to a normally harmless food or ingredient. The body responds to the substance's presence by producing antibodies. These trigger the release of histamine and other chemicals into the bloodstream, resulting in an allergic reaction. But the two schools differ over the type of symptoms that such allergic reactions can cause.

The rice alternative
If you are allergic to or are unable to tolerate gluten in wheat, try substituting rice-based for wheat-based products—from boiled rice to use in cakes and breakfast cereals.

How to stop food allergies from developing

Breastfeeding for at least four months, along with the gradual introduction of new foods, can help to reduce the risk of food allergies. For example, delay giving allergy-provoking foods such as wheat, citrus, dairy products, eggs, nuts, soy, and fish until after six months. Introduce one new food at a time, so that you can see if a particular food causes a reaction.

If you think your baby is allergic to a food, the next step may be to try excluding it from his or her diet. Before you do this, however, seek professional advice from a pediatrician, healthcare provider or dietitian.

Immediate food allergy

According to conventional medicine, food allergy reactions are responsible for only a narrow range of symptoms that come on very quickly after the trigger food is encountered. The typical signs are swelling of the lips, urticaria (nettle rash), vomiting, and diarrhea. In severe reactions, a condition called anaphylactic shock

Some foods contain such potent allergens that you don't have to eat them to get symptoms. Steam from cooking fish can trigger an asthma attack in susceptible people and even kissing someone who has eaten fish can cause problems.

can cause breathing difficulties, a sudden drop in blood pressure, and, in some cases, death.

Hidden food allergy

Adherents to the hidden food allergy theory believe that food molecules are constantly leaking through our intestinal walls into the bloodstream, triggering low-intensity but pernicious immune reactions that result in chronic illnesses, typically including digestive problems such as diarrhea, constipation, flatulence, irritable bowel syndrome, and even diverticular disease. These reactions may not appear for hours or even days after the trigger food was eaten.

Intolerance and sensitivity

Food intolerance or sensitivity occurs when the body cannot digest a particular food. One classic example is lactose intolerance, where the sufferer does not produce the lactase enzyme needed to break down lactose, a major component of milk. Another common type of intolerance is celiac disease. The intestines of people with this condition cannot deal with and absorb gluten, a protein found in wheat and some other grains.

Alternatively, sensitivity may be the direct result of the pharmacological effect of chemicals present in a food, such as caffeine in coffee or amines in cheese and chocolate. Food intolerance does not involve the release of antibodies but can cause many symptoms that alternative allergists link to hidden food allergies, such as flatulence, headaches, or fatigue.

BEATING FOOD REACTIONS

You can take a number of steps to deal with the problems caused by food reactions, but the golden rule is "avoidance." The only sure way to avoid developing a reaction to a food or food ingredient is not to eat it. First, however, you need to identify what it is you're eating that is causing the problem.

Identifying food triggers

Immediate food allergy triggers are often easy to recognize because they produce symptoms soon after being eaten. If you develop a rash, swelling, nausea, shortness of breath, or itchiness of the lips, tongue, and throat over a short period of time, you should suspect whatever you just ate. Foods that most often trigger this type of reaction are peanuts, nuts, shellfish, and eggs, although practically anything can cause a reaction in a susceptible individual.

Steering clear of a food that has triggered an immediate allergic reaction is essential, because subsequent reactions may be a great deal

Common allergy-provoking foods

Several foods and food groups are common triggers for food intolerance or allergy. Some foods are associated with mild reactions; others can cause life-threatening symptoms.

Source	Reactions
COW'S MILK	A common cause of allergic reactions in children causing vomiting, colic, and diarrhea. In 9 out of 10 cases, sensitivity disappears by the age of 3.
EGGS	Eggs are common allergy-provoking foods in children, causing rashes and stomach upsets, but sensitivity to them gradually disappears with age.
FISH	Fish can produce severe allergic reactions within minutes of being eaten; they can cause rashes, asthma, vomiting, diarrhea, and anaphylaxis.
WHEAT	Wheat, rye, and barley usually cause gastrointestinal symptoms, including diarrhea. Wheat flour can provoke asthma.
PEANUTS	One of the most potent allergens, even a trace amount of peanut can cause life-threatening anaphylaxis in susceptible individuals.
SHELLFISH	Shrimp, lobsters, crabs, and other shellfish can cause violent allergic reactions, such as urticaria (nettle rash) and swelling of the lips.

stronger. You could even be in danger of a life-threatening attack of anaphylaxis. People who know that they are at risk for extreme allergic reactions, whether to peanuts or bee stings, should always carry an epinephrine self-injection kit with them just in case they are exposed. You should check with your doctor if you are in any doubt.

The date, the time, the food...
Keeping a food diary can help you to identify possible food allergy triggers. Note everything you eat for a week; include any symptoms, such as headaches, rashes, and stomach upsets.

Hidden food allergy triggers may be less dangerous, but they are also much harder to identify. One of the reasons that conventional medicine is unwilling to accept hidden food allergy as a genuine phenomenon is that it is so hard to pin down. Reactions may not occur until days after a food was eaten, involving apparently unrelated symptoms of varying severity that actually get worse when the trigger is avoided, because of a kind of withdrawal reaction. Even more confusing, some reactions happen only when two or more foods are eaten in combination.

If you are experiencing a number of vague, chronic, and apparently unrelated symptoms, get them checked out by your doctor. If no conventional explanation can be found, you should consider the possibility that you might have a food allergy.

Keeping a food diary

A reliable aid in identifying a hidden food trigger is to keep a food diary. Record the type and severity of your symptoms on a daily basis. At the same time, keep a record of everything that you eat or drink, including all the ingredients. It is important to be comprehensive because even minute quantities of a trigger substance could be responsible for your symptoms.

You can use a food diary to identify possible foods and then try to eliminate them from your diet. If a week's abstinence produces results, you can confirm your suspicions by eating a little of the suspect—if it is the trigger, symptoms should reappear, possibly quite strongly, within a couple of days. People with special dietary requirements, such as children, the elderly, pregnant women, and vegetarians, should never restrict their diets without checking with a doctor first.

LACTOSE INTOLERANCE

Lactose intolerance is a condition in which the body fails to produce the lactase enzyme. Lactase is necessary for the digestion of lactose, a natural sugar present in milk. Without it, lactose cannot be broken down and passes unchanged into the large intestine where it is fermented by intestinal bacteria, causing bloating, abdominal pains, and diarrhea.

Most ethnic groups in the world are lactose intolerant; some would say this is the "normal" adult state, because we're not designed to drink milk after being weaned. In countries where dairy products are not regularly eaten, the body stops producing lactase at some point between infancy and adulthood. In the Western world, however, we include so many dairy products in our diet that we need to produce lactase to deal with the milk sugar.

positive health tips

Decoding ingredient labels

Many processed foods contain lactose, cow's milk protein, wheat, or gluten. If you need to avoid any of these foods, check the list of ingredients on the label to see if they include any of the following.

- **Gluten-free diet** Watch out for barley, bran, cereal binder, cereal filler, starch, cereal protein, modified starch, edible starch, food starch, flour, rusk, rye, vegetable protein, wheat flour, and wheat germ.

- **Dairy-free diet** Steer clear of milk protein, milk powder, animal fat, casein, caseinate, hydrolyzed casein or whey, lactose, non-fat milk solids, lactalbumin, lactoglobulin, skim milk powder, and whey.

Lactose-free eating

Foods to avoid	Substitutes
Cow's milk (including condensed milk, and evaporated milk), goat's milk, sheep's milk, and buttermilk	Soy milk, rice milk, coconut milk, oat milk, almond milk, and 95% lactose-free milk
Butter or margarine	Milk-free margarine—there are several brands to choose from, including some made from olive oil
Cheese (including cream cheese and cottage cheese)	Dairy-free cheeses, such as soy, are available in both soft and hard varieties
Yogurt	Dairy-free yogurts—there are lots to choose from
Ice cream	Sorbet or soft nondairy frozen yogurt

Lactase deficiency can also occur as a rare inherited condition that becomes apparent shortly after birth. Much more common is temporary lactose intolerance, which can occur after an upset stomach.

People with lactose intolerance should avoid milk, including sheep's and goat's milk, and dairy products such as cream, yogurt, cheese, and butter. Processed foods, monosodium glutamate, artificial sweeteners, and medicines may also contain lactose.

Some people with lactose intolerance can handle small amounts of milk and dairy foods, such as milk in tea or coffee or the occasional cheese sandwich, without experiencing any ill effects.

For those people who cannot tolerate even small quantities of dairy produce, most health food shops and some supermarkets stock low-lactose and dairy-free

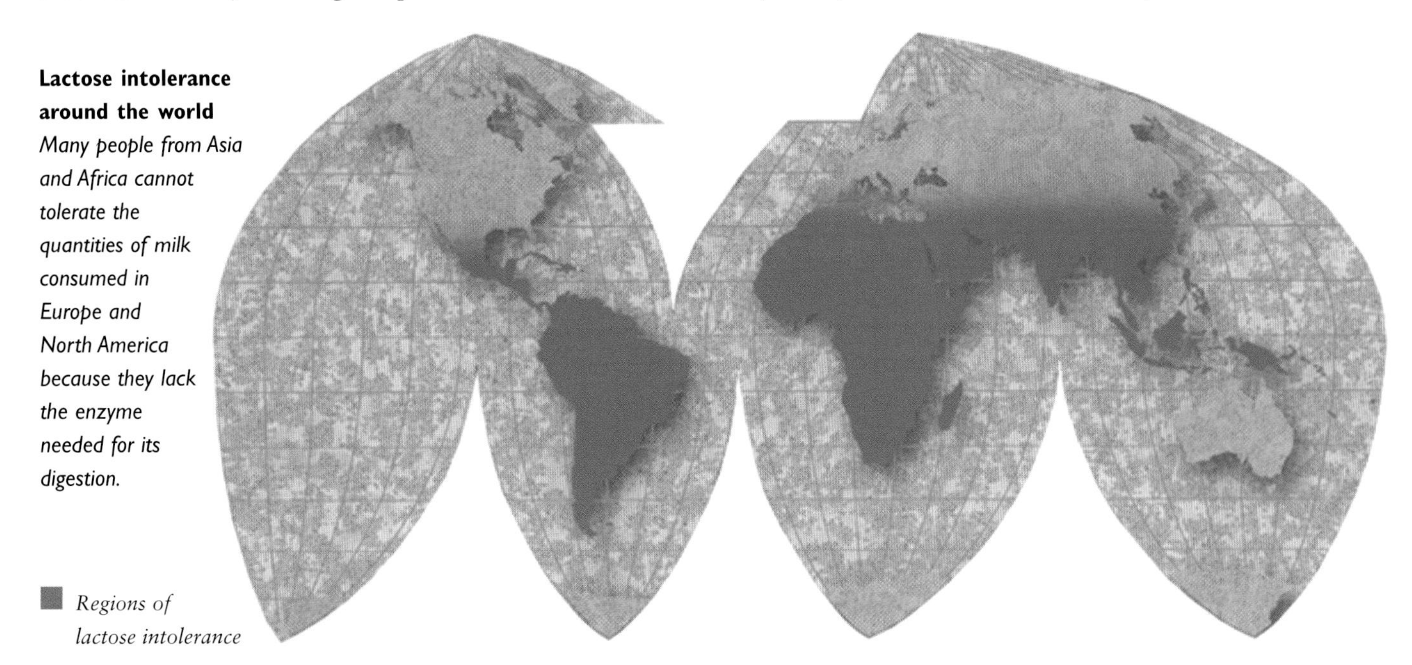

Lactose intolerance around the world
Many people from Asia and Africa cannot tolerate the quantities of milk consumed in Europe and North America because they lack the enzyme needed for its digestion.

■ *Regions of lactose intolerance*

alternatives, such as cream substitutes. If you have to limit your intake of milk and dairy products, you need to ensure that you get enough dietary calcium from other sources.

GREAT GLUTEN-FREE MEALS

For most people, wheat is an essential part of the diet, the basis of such staples as bread and pasta and an ingredient in countless other foods. But for people with celiac disease, the gluten protein found in wheat, rye, oats, and barley causes serious health problems. Even minute amounts of gluten can cause the intestinal tract to become inflamed and damaged—the result is abdominal pain and diarrhea. Also, important nutrients aren't properly absorbed.

A growing number of people claim that wheat allergy is a contributing factor in some health complaints, such as arthritis. Although there is little evidence that these people actually have true allergic reactions, eliminating wheat from the diet sometimes seems to help.

A gluten-free or wheat-free diet means eliminating all wheat, barley, and rye and products made out of grain, such as pasta, bread, breakfast cereals, and cookies. But it's not always easy to identify products that contain gluten. Many manufactured foods contain flour as a thickening agent, so read the labels on products to double check.

Gluten-free eating

Foods to avoid	Substitutes
Wheat, oats, barley, rye	Rice, potatoes, soy, corn (maize), chickpeas, polenta (cornmeal), millet, buckwheat, tapioca, and sago
Meal or flour made from wheat (including white, whole-grain, or self-rising), barley, rye, or oats	Flours made from nonwheat sources: rice, potato, soy, corn (maize), chickpea, arrowroot, sago, or tapioca
Bread and any dish that includes breadcrumbs	Gluten-free bread or bread mix, rice cakes, rice crackers, or taco shells; use crushed cornflakes for breadcrumbs
Breakfast cereals containing wheat, bran, oats, or barley	Cornflakes, puffed rice cereal
Ordinary pasta	Rice or buckwheat noodles, gluten-free pasta
Cakes, cookies, crackers, pastry, batter, semolina, and couscous	Specialty wheat/barley/rye/oat-free products
Sauces and gravies thickened with flour	Oil, homemade dressings; homemade gravies thickened with cornflour; butter or margarine
Malt vinegar	Wine or cider vinegar

EXERCISE AND DIGESTION

Your digestive system is surrounded and supported by a complex set of abdominal muscles, which play a vital part in the process of digestion. Just like all the other muscles in your body, these need attention to ensure that they continue to function properly. Here we show you exercises that help to maintain abdominal muscles in good working order. What's best before a workout—a banana or an energy bar? Discover how to best fuel your body for exercise, as well as when and how to restore your energy afterward.

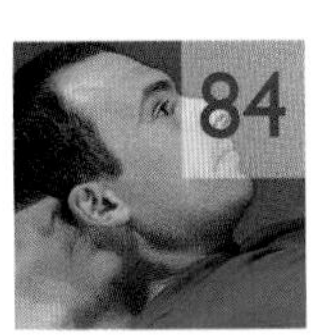

Exercise and your digestive health

You probably recognize the importance of regular physical activity for your heart and lungs, but did you know that the digestive system benefits from exercise? Regularly working out helps to keep your GI system in good working order.

THE SUPPORT SYSTEM

Your abdominal muscles are part of the set of structures that keep your digestive organs in their proper places. The muscles form a complex web that extends from the diaphragm (which forms the roof of the abdominal cavity) to the muscles situated in the pelvis (which form its floor). Abdominal muscles also play a vital role in maintaining posture and producing twisting, turning, and bending movements of the torso.

In addition to helping support your intestines, the muscles of the pelvic floor have a direct effect on your large intestine. As they contract and relax, they squeeze the colon (the main part of the large intestine), and help to move the contents along so that they can be eliminated.

Shaping up the GI system

It's difficult to exercise the digestive system directly, but keeping your abdominal muscles in shape not only helps to keep you toned but also to keep your system moving.

The walls of your intestines are partly composed of muscle fibers, allowing the intestines to contract and move food along by peristalsis. Intestinal muscle fibers are a special kind, known as involuntary—unlike the muscles in your arms or legs, they are not under conscious control.

WORKING OUT

By being moderately active for 30 minutes most days of the week (taking a brisk walk, for example), you can improve your GI motility—the amount and efficiency of peristalsis in your intestines. (The USDA now recommends one hour of exercise each day.) When you exercise regularly, you indirectly affect the digestive smooth muscles as your abdominal muscles contract and relax. This affects GI motility, helping food to move through your GI system at the right speed. Lack of activity leaves your digestive system sluggish. Sedentary people will often suffer from constipation and increase their risk of developing diverticular disease.

Can massage improve GI function?

Massaging your abdominal muscles can increase the blood supply to your GI system, aiding the movement of food along your digestive tract, and it is said to offer relief from indigestion and stomach tension. Massage in a clockwise direction to match the flow of the digestive tract. Use gentle strokes that work from the right to the left side of your body.

ASK THE EXPERT

TURNING UP THE THERMOSTAT

Your metabolism (the rate at which you burn up energy) is set by an internal thermostat. Being able to "turn up" this thermostat would make you burn energy faster, using up carbohydrates and fats before they get laid down as fat and even consuming preexisting fat deposits. One of the best ways to reset your metabolic thermostat is through moderate exercise.

Eating and exercising

Your body needs fuel to exercise, but eating too much before a session will cause a tug-of-war between your digestive system and your muscles. Time your eating to supply the energy your body needs without upsetting your digestion.

To exercise, you need energy. This energy comes from the breakdown of simple carbohydrates, which the blood takes to the muscles ready to be used. Energy stores are also found, in limited quantities, in the muscles themselves (in the form of a substance called glycogen).

Carbohydrates come from the foods you eat, and one of the main jobs of your digestive system is to break down these foods so that they can be absorbed into the bloodstream. In order to exercise safely and effectively, you need to maintain the level of carbohydrates available to your muscles by supplying the right things to your digestive system.

FUELING YOUR EXERCISE

U.S.D.A. guidelines for a normal balanced diet for general health recommend that we obtain most of our calories from complex carbohydrates, such as bread, potatoes, pasta, legumes, fruit and vegetables (11–20 servings per day); that we limit our intake of Calories from fats, avoiding saturated fats where possible, and choosing low-fat milk, cheeses, and spreads; and that we eat 2–3 servings of proteins (lean meat, fish, eggs, and legumes) and 2–3 servings of dairy products. In summary, the largest part of your diet should contain fiber and complex carbohydrates, and only a small proportion should contain fats and simple carbohydrates (see pages 62–65 for more information).

FEEDING YOUR MUSCLES

These eating guidelines help to prepare you for exercise by improving your general health. If you are getting moderate exercise every day, you may also need to further boost your carbohydrate intake in order to compensate for the increased amount of energy being used by the muscles.

Your muscles quickly use up their stores of glycogen, especially during exercise. These stores need to be built up before exercise, and replenished during and after exercise. Only about 600–800 Calories worth of glycogen are actually stored in the body at any particular time (mostly in the muscles, but there is also some in the liver). If you allow your glycogen stores to become exhausted, your athletic performance will be impaired.

Eating after a workout
A light snack after exercising can replenish your muscles' depleted energy stores. A banana supplies about the right amount of carbohydrates.

HIGH GLYCEMIC INDEX FOODS

These foods produce a rapid rise in blood glucose; they are useful after exercise to boost glycogen stores.

- wholemeal bread
- cantaloupe melon
- diluted orange juice
- carrots
- processed breakfast cereals
- raisins

LOW GLYCEMIC INDEX FOODS

These are converted to sugar slowly so they are useful between workouts or a few hours before exercise.

- cherries
- skim milk
- apples
- dried apricots
- wholewheat pasta
- lentils

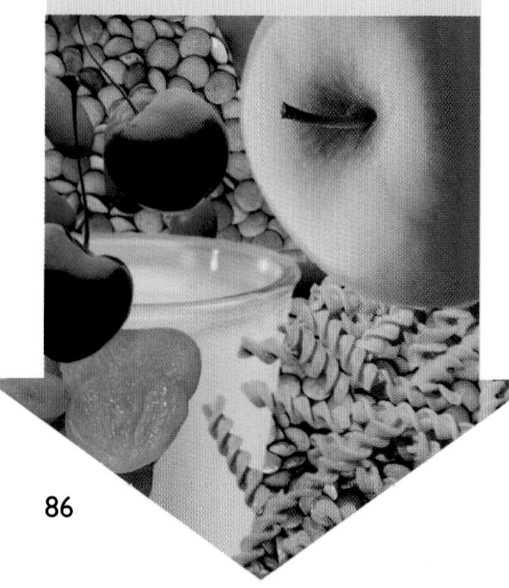

Meeting extra demands

In order to meet increased calorie demands and build up and maintain stores of glycogen, you will need to increase your energy intake. The amount you need to eat depends on the intensity of your exercise program. If you get regular exercise for general health maintenance, for example swimming or jogging for 30 minutes five days per week, you probably need to only boost your food intake a little.

These increases should come from eating more complex carbohydrates, which should account for a higher proportion of your diet—around 65 percent—than in the normal diet. You should avoid eating more fats or simple sugars than you already do, except possibly during exercise or immediately after exercising, when you need to replenish glycogen stores quickly. Active people also need more protein per pound of body weight than their couch potato cousins. Between .5 and .77 grams per pound of body weight should be enough to promote healthy new muscle growth and for tissue repair.

The glycemic index

How do you know whether a food is a good source of complex carbohydrates or is rich in simple carbohydrates? A good guide that also provides useful information for the diet-conscious exerciser is the glycemic index of a food. This index is a measure of the rate at which consumption of a food raises the level of blood glucose. In other words, it shows how quickly a food gets broken down by the digestive system, converted into glucose and absorbed by the blood. The fastest carbohydrate to be absorbed into the bloodstream is pure glucose and this is given an index rating of 100.

Foods that have a low index rating and are broken down slowly make you feel full for longer and keep blood sugar levels constant. This helps to prevent you from craving more Calories than you need and reduces the degree to which carbohydrates in your blood are converted into fat. Low glycemic index foods include beans and legumes, apples, wholewheat cereals, dried apricots, pasta, and oats. Foods with a high glycemic index include bread, glucose energy drinks, sugar, honey, mashed potatoes, and breakfast cereals.

But how does this help you plan a diet for exercise? The best way of ensuring a good level of carbohydrate intake in a healthy form is to eat foods that have a low glycemic index—in other words, complex carbohydrates. This will ensure that your body is fueled up and ready for exercise, with full stocks of glycogen. Immediately after a workout, when your muscle glycogen is depleted, you need to get your blood sugar levels up quickly. This is easily achieved by having high glycemic index foods or drinks, such as rice cakes and jelly or a high-energy drink, in the two hours after exercise.

If glycogen is not replaced within two hours, it can take up to 48 hours for glycogen stores to refill. Glycogen is replaced quickest in the first 30 minutes after activity.

THE BATTLE FOR BLOOD—MUSCLES VERSUS DIGESTION

An important consideration in terms of meal timing and exercise is the issue of blood flow to the different areas of your body. You only have about 8 pints of blood to supply all of your body's needs. During exercise, your heart rate increases two or threefold, massively boosting the volume of blood pumped per minute (your muscles get up to 30 times more blood per minute than when you are at rest). Putting food into the digestive system before or during exercise can lead to conflicting demands, because your stomach and intestines compete with your muscles for limited resources.

Athletes in regular training need at least 5,000 Calories a day, compared with the average daily requirement of around 2,000–2,500 Calories for most adults.

Digestive needs

After a meal, your body begins the process of digestion and there is a greater need for blood in the GI system so that nutrients can be absorbed and carried to the liver, and then on to the different parts of the body. The conversion and use of these nutrients becomes a priority to your body, and in order to transport them to your body tissues, the blood vessels in and around muscle tissues constrict while vessels supplying your digestive organs dilate. This directs blood away from muscles towards the digestive system, reducing the provision of oxygen and nutrients to the muscles. Assuming you are not too active after a meal, this is not a problem for your body.

Blood for working muscles

If you start exercising soon after eating a meal, your central nervous system attempts to redirect blood flow away from your digestive organs and toward the working muscles. This redirection of blood flow slows the digestive processes and reduces GI motility. A sort of "tug-of-war" between your muscles and digestion results, and both your digestive and your athletic performances are compromised. You may experience rapid breathing, an abnormally high heart rate, nausea, and dizziness during or after your workout. If these symptoms occur, you should slow down or stop your workout altogether.

For serious athletes, sports nutritionists recommend eating five or six smaller meals throughout the day, thereby ensuring that there is sufficient glycogen and blood-borne glucose in the system to supply the energy needed for an efficient workout, without straining the digestive organs.

positive health tips

Remember the fluid factor

The simple steps given below outline how you can avoid dehydration.

- Ensure that you are fully hydrated before beginning your workout. This does not mean downing a quart of water before you go running, although it's wise to have a glass. Instead, ensure that you keep up your fluid consumption throughout the day and don't let yourself get thirsty.
- Start drinking early in your exercise program—don't wait until you are thirsty. Take frequent small sips. A cyclist's water bottle is a good container.
- Put a teaspoon of sugar in your water bottle—a low-sugar concentration increases the rate at which fluid is absorbed by your digestive system and gets into your bloodstream.
- Take plenty of fluids on completion of your exercise program. Ideally you should carry your own supply of fluid in your gym bag.
- Use isotonic or energy drinks with care. They can be useful for fluid replenishment and also provide additional carbohydrates. If used incorrectly, however, they may lead to nausea and stomach cramps and can actually inhibit fluid absorption.

MAINTAINING FLUID LEVELS

During exercise your muscles are burning fuel to produce energy at a tremendous rate. Unfortunately, the system is not particularly efficient—only about 25 percent of the energy stored in our food is converted to a

EATING AND DRINKING PLAN FOR MODERATE EXERCISE

It's important to get the timing of food intake right when exercising. Drink plenty of fluids to prevent dehydration. Avoid exercising within two hours of eating a heavy meal.

Food and fluid before activity

One hour before exercise, have a snack that is high in complex carbohydrates, such as a small sandwich or banana. Have a large glass of water or diluted fruit juice just before exercising.

Food and fluid during activity

There is no specific need to eat if you are exercising for less than one hour; if you are exercising for longer than an hour, have small amounts of fruit or diluted fruit juice.

Food and fluid after activity

About 30–60 minutes following exercise eat some fruit, a small sandwich, or a whole-grain muffin. Drink a large glass of water or diluted fruit juice.

form that the muscles can use to contract and generate force. Most of the rest becomes heat, and your body needs to dissipate this excess heat. An effective way of doing this is through sweating.

In addition to the water lost through sweating, heat-laden moisture is also lost every time you breathe out. This means that during a prolonged workout, you could very easily become dehydrated. This will impair your athletic performance more severely than a lack of glucose would. In other words, correct hydration (maintenance of fluid levels) is more important than nutrition.

The factors that influence your exact fluid requirements include the length of the workout, the intensity of the exercise and also the temperature in the environment (outside or inside). If you are just jogging for half an hour, there's no need to be concerned about dehydration, but it's good to get into the habit of having a large glass of water before any activity.

TIMING MEALS

The best time to eat and drink, in relation to exercise, depends on a range of factors—the glycemic index of the foods you're eating (see page 86), fluid requirements, and blood flow issues. In addition, each person is different in his or her eating habits and in his or her response to food, so it is important to listen to your body and develop a routine that works for you.

Putting all the advice together, however, we can derive some general guidelines to get the best out of any exercise regimen (see above).

Exercising on an empty stomach

Choosing when to exercise is often a matter of individual preference, circumstances, and opportunities. Some people like to snack an hour beforehand, while others find that they cannot tolerate anything in their stomach. Exercising on an empty stomach is quite safe. During activity, fluid replacement is much more important than eating.

PREVENTING A HERNIA

If the muscles of your abdomen are weak, a sudden increase in pressure may cause a hernia—the protrusion of a loop of intestine through the abdominal wall. When working out with weights, do not try to lift too much or you could strain and give yourself a hernia. Sensible breathing and avoiding lifting more than you

EATING AND DRINKING PLAN FOR INTENSE EXERCISE

The correct timing of food intake is particularly important when you do strenuous or endurance exercise. Drinking plenty is important to prevent dehydration. Avoid exercising within two hours of eating a heavy meal.

Food and fluid before activity

About 15–30 minutes before exercise have a snack that supplies about 2 ounces of carbohydrates—for example a handful of dried fruit, a 2-ounce bowl of cereal, or two to three bananas.

Food and fluid during activity

Consider having an isotonic drink to supply both fluid and glucose.

Food and fluid after activity

Thirty minutes after exercise, eat some fruit (especially bananas) or a baked potato with beans, and drink at least a pint of water. These are vital for replacing glycogen stores.

can realistically manage will keep your abdominal wall intact and prevent a hernia.

Abdominal breathing

The breathing pattern of the average sedentary person is one of small, shallow breaths, which use only the top part of the lungs. As a result, the lungs are never stretched to capacity—and the diaphragm is rarely exercised. Correct abdominal breathing allows you to fill your lungs from the bottom to the top. By breathing in this way, you extend and exercise the diaphragm, keeping it in good shape.

Exercising your diaphragm

Try this simple exercise to improve your abdominal breathing. Place one hand on your chest and one hand on your abdomen and take a deep breath. If you feel your chest rise first, you are filling only the top half of your lungs and are failing to exercise your diaphragm. Aim to get your abdomen rising before your chest rises. This not only benefits your diaphragm, but also ensures that you are using all of your lung capacity. As you breath out, concentrate on emptying the lungs and allowing your abdomen to push out. Abdominal breathing is a basic yoga technique: in addition to its beneficial effect on the diaphragm, yoga can also help to relieve stress.

EXERCISE FOR DIGESTIVE PROBLEMS

A number of GI disorders can make physical activity difficult, yet people with impaired physical health usually benefit the most from exercise. If you suffer from some form of digestive condition, staying active is one of the most important steps you can take to help your GI system stay healthy and to prevent further problems. Whatever exercise you choose, it is important that you enjoy it and that you feel comfortable taking part.

Stay fit without straining

If you have a hernia or a colostomy (a surgically created opening from the intestines to the abdomen), aim for 30 minutes of moderate activity on most days of the week. Walking is an excellent way to stay fit and healthy and it is a good way for many people with digestive problems to remain fit without putting undue strain on the digestive system.

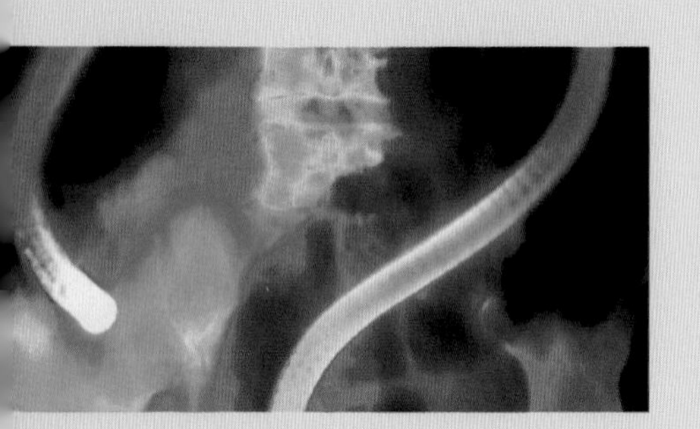

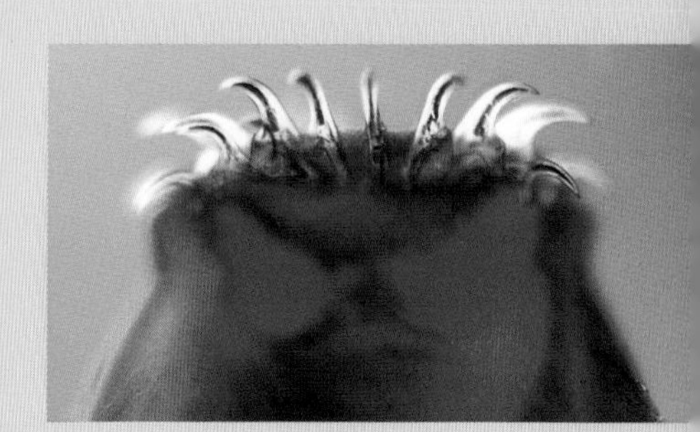

3 What happens when things go wrong

Knowing what can go wrong

Digestive symptoms are extremely common and usually pretty harmless, but as we grow older the risk of developing a more serious digestive disorder increases, so greater vigilance is called for. Here we outline the main factors that cause or contribute to digestive complaints.

INFECTION

On a worldwide basis, infection is by far the most common cause of gastrointestinal and liver disease. Diarrheal diseases, for instance, are extremely common. In developed countries, diarrhea is usually a short-lived problem, but in the developing world it kills about 3–4 million people each year, mostly preschool children.

Helicobacter pylori infection is probably the most common in the world. Many people live in harmony with this bacterial organism without it causing any disease; in some susceptible individuals, however, it causes peptic ulcers, whereas in others it can contribute over many years to the development of stomach cancer. Research efforts currently focus on understanding why some people with this infection suffer no ill effects, and indeed may be protected from other infections. As with many conditions, a person's genetic makeup may be the key.

Stomach ulcers are more likely to afflict people in blood group A, whereas duodenal ulcers mainly affect people in group O.

Other causes of infection are parasites, such as liver flukes and tapeworms, which today occur mainly in developing countries where sanitation is poor.

GENETICS AND HEREDITARY FACTORS

Inherited (genetic) factors play an important role in some common digestive disorders, although in most cases an inherited tendency is not the only factor required to develop the disease. The most common example is colon cancer, where genetic predisposition acts in a number of different ways (see page 97). For instance, in very rare cases, members of a family carry an abnormal gene that predisposes them to develop colon cancer, conditions known as familial adenomatous polyposis or hereditary nonpolyposis colon cancer. These types of cancers account for about 1 percent of colon cancer cases. In such families, children of an affected person carry a 1 in 2 chance of inheriting the disorder. Screening programs are now available so that people at risk ican be detected and preventative steps taken to stop any cancer from developing.

Such straightforward genetic effects are rare, but even in the other 99 percent of colon cancer patients, family history affects the risk status of an individual. Other digestive disorders that may run in families include the inflammatory bowel diseases Crohn's disease and ulcerative colitis, and celiac disease.

AGING PROCESSES

In general terms, the digestive system can take a lot of punishment without significant loss of function, so that the effects of aging do not usually result in serious damage. But there are more specific changes that may produce visible effects. For instance, advancing years result in metabolic changes, including decreased physical activity, increase in body fat, and decrease in muscle bulk.

MORE COMMON

HEMORRHOIDS
PRESENT IN UP TO HALF THE POPULATION BY AGE 50

IRRITABLE BOWEL SYNDROME
AFFECTS 15% OF THE U.S. POPULATION

PEPTIC ULCER
CHANCE OF DEVELOPING OVER LIFETIME IS 1 IN 10
Contributing factors include aspirin and other NSAIDs, the bacteria Helicobacter pylori, *cigarette smoking, alcohol abuse, and stress.*

CELIAC DISEASE
AFFECTS 1–50 TO 1–250

CIRRHOSIS
3,000,000 PER YEAR
Kills 25,000 each year. Eighth leading cause of death by disease.

CROHN'S DISEASE
400,000–500,000

COLORECTAL CANCER
148,000 NEW CASES EACH YEAR
The second most common cause of cancer-related death in the United States

As people age, they produce less acid in their stomachs, probably because of a long-standing infection by *Helicobacter pylori*. Reduced levels of stomach acid can interfere with the processes of digestion, which can make it more difficult to absorb certain vitamins and minerals, such as iron, folic acid, vitamin B_{12}, calcium, and vitamin K. Some age-related digestive health issues are linked to the fact that older people typically have a poorer diet. The reasons for this are complex, but they may include poor general health and mobility, financial difficulties, and state of mind, as well as physiological changes such as changing senses of smell and taste with age.

FUNCTIONAL DISTURBANCES

The brain and gastrointestinal (GI) system are intimately connected: We're all familiar with terms such as "butterflies in the stomach" or "gut feelings" that relate to our state of mind. The GI system has its own branch of the peripheral nervous system—the enteric nervous system—which is extensive and regulates digestion. It is not surprising, therefore, that the GI system can reflect states of mind, such as emotional upsets or depression. When investigating and treating functional disturbances, doctors often work closely with psychologists to identify and address possible causes.

The enteric nervous system can play a more specific role in gastrointestinal problems. Some doctors believe that events affecting the bowel, such as attacks of gastroenteritis, can sometimes cause the intestine to become more sensitive, resulting in abdominal symptoms. This disturbance of bowel function is not strictly a disease, but it is a real physical phenomenon, which doctors sometimes refer to as a functional disturbance. The most commonly recognized form of functional disturbance is irritable bowel syndrome.

TUMORS

Cancer of the digestive system is very rare in young people but becomes more common with age, most notably after 50. As elsewhere in the body, tumors in the GI system may be benign or malignant. Stomach cancer is the second most common cancer worldwide, after lung cancer. It has a particularly high rate of incidence in Japan, where it's thought to be linked with the high consumption of smoked, salty, and pickled foods. Doctors now know a great deal about tumors affecting

Influences on digestive health
Throughout your lifetime, your digestive system has to cope with a variety of factors that can harm it. Some of the diseases that can result, as well as how they affect the U.S. population, are shown below.

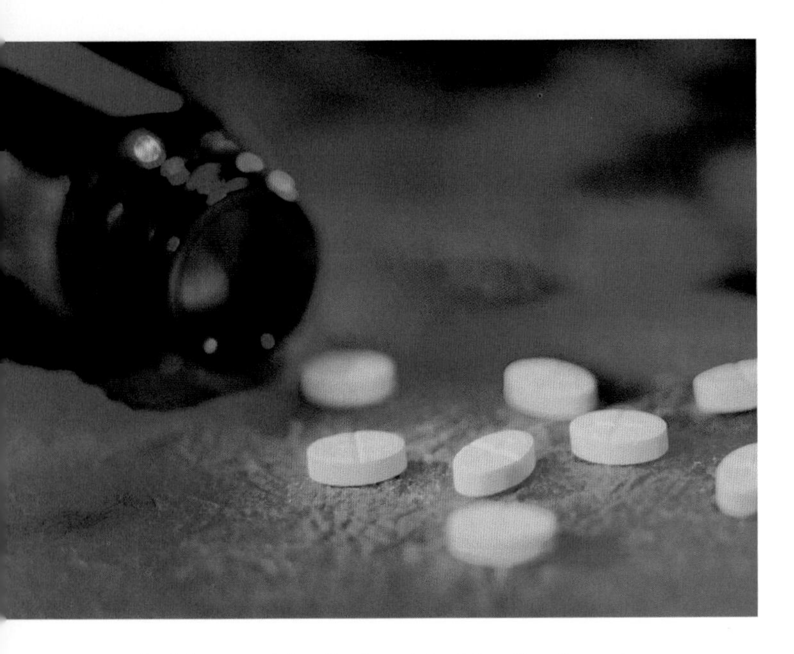

Aspirin—good for the heart, bad for the stomach
The group of drugs collectively known as nonsteroidal antiinflammatories (NSAIDs) can have a harmful effect on the stomach lining, in some cases causing ulcers when used for a long time.

the colon, in particular, and the molecular processes involved in the formation of benign growths (polyps) and cancer. What is less certain is the role of external factors, such as diet, that may trigger the process.

LIFESTYLE—ALCOHOL, EXERCISE, DRUGS

Chronic heavy alcohol use can lead to many health problems. Liver damage is probably the best known, and typically follows one of three patterns: fatty liver, alcoholic hepatitis (inflammation of the liver cells), and cirrhosis of the liver. Alcohol can also cause a condition known as pancreatitis (inflammation of the pancreas).

Physical exercise keeps the bowels regular. In an increasingly sedentary Western lifestyle, more and more people suffer from constipation.

Prescription drugs often have an impact on the digestive tract. A common side effect of antibiotics, for example, is diarrhea, whereas strong painkillers, such as codeine and morphine, can cause constipation. Aspirin and related drugs can cause stomach upsets, whereas long-term use can cause ulceration and worsen conditions such as existing peptic ulcers or inflammatory bowel disease.

Intestinal parasites
According to one theory, lack of intestinal parasites, because of improved hygiene, may be a factor in the rise of autoimmune diseases of the digestive system.

INFLAMMATION

There are a number of conditions in which the body's immunological response is overstimulated, causing inflammation when there is no detectable infective agent, such as a bacterium, virus, or parasite. Doctors believe that these autoimmune diseases may be triggered by childhood infections or imbalance in gastrointestinal flora.

Inflammatory bowel disease is one such condition, manifesting itself in the form of ulcerative colitis (inflammation of the colon) or Crohn's disease (inflammation of any part of the gastrointestinal tract). Research shows that this class of diseases is increasingly common; some experts put this down to our "superhygienic" lifestyle, which prevents any contact between children and "bugs"—specifically certain types of bacteria or intestinal parasites such as worms.

MOTILITY PROBLEMS

The digestive tract is supplied by a complex nervous system that coordinates the contraction of the gastrointestinal muscles and therefore the movement of food through the digestive system. In a number of rare conditions, these mechanisms do not work properly. In achalasia, for example, which has an incidence of just 1 per 10,000, the muscle at the lower end of the esophagus does not relax normally in response to swallowing food, making it difficult to swallow.

CONGENITAL ABNORMALITIES

Occasionally, parts of the digestive system fail to develop properly in a fetus and the baby is born with a congenital defect. The most common congenital problems are cleft lips and palates. In the U.S., about 1 in 700–750 babies is born with a cleft lip or a cleft palate. Today, these conditions can be successfully treated with plastic surgery while the baby is young (from 3 to 15 months).

Who's who—stomach and digestive system experts

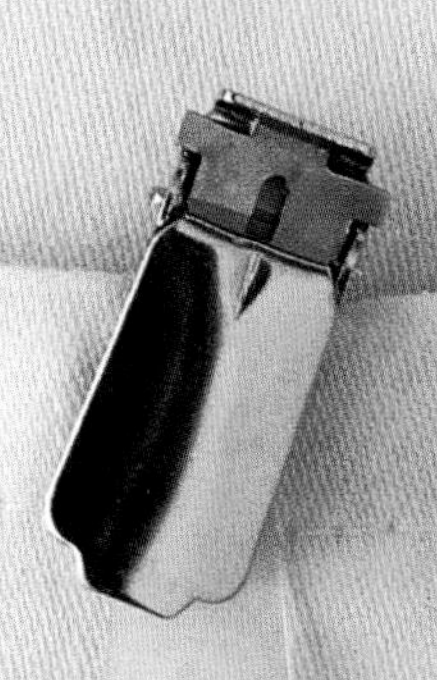

Digestive disorders are extremely common, and unlike many diseases of the heart or brain, they can often be dealt with very effectively by your physician without the need for a referral to a specialist.

PERSONAL PHYSICIAN

The personal physician is in the best position to give a preliminary assessment of someone's symptoms and to judge according to national guidelines whether they are serious enough to refer the patient to hospital. After carrying out a physical examination, the doctor can treat most disorders. Some doctorss may also take a more active part in the gastroenterology team and some may even perform endoscopies (looking inside the digestive tract) themselves.

GASTROENTEROLOGICAL SURGEON

Working closely with the gastroenterologist, the surgeon has special expertise in digestive diseases that can be treated surgically. Increasingly, surgeons will specialize even further, often concentrating on a particular part of the digestive system such as the colon and rectum or the liver and gallbladder. Common operations in this area are the removal of the gallbladder, hernia repair, and appendectomy (removal of the appendix).

PATHOLOGIST

Not directly involved in dealing with patients, pathologists provide a support service to doctors and surgeons, assessing tissue specimens taken during surgery, as well as biopsies from endoscopic procedures. The pathologist is often the person who makes the final, firm diagnosis and thus plays a vital role in the team.

NURSES

Many hospitals have nurses specializing in nutrition and gastroenterology, who are experienced in treating patients with certain problems, such as those requiring artificial nutritional support, people with inflammatory bowel disease, or those being treated for viral hepatitis. Nurses are also key to the running of an endoscopy unit. They admit the patient on arrival, assist with the endoscopy, look after equipment, and care for the person undergoing the procedure. Some nurses undergo further training and perform some endoscopic examinations themselves.

GASTROENTEROLOGIST

The gastroenterologist is a hospital-based doctor who specializes in the diagnosis and treatment of diseases of the digestive system. The gastroenterologist will deal with most forms of gastrointestinal disease as well as diseases of the liver and pancreas. Most gastroenterologists perform endoscopies for further diagnosis and therapy.

DIETITIAN

Most gastroenterology teams will have a dietitian (nutritionist) who is trained to provide specialist advice on diet and nutrition. This may take the form of directly advising patients on dietary supplements and special requirements such as gluten-free diets for people with celiac disease. They also work in close liaison with the medical team to provide nutritional support for poorly nourished patients; for patients requiring a liquid diet via a tube into the stomach, such as burn victims or those with other traumatic injuries; and for people who are unable to eat and therefore need intravenous feeding.

CLINICAL PSYCHOLOGIST

Digestive disorders may have a component that is related to stress and other emotional factors. In some units, therefore, a clinical psychologist may be employed as part of the team. A psychologist has training in theories of human behavior and thought and possesses special counseling skills. These techniques can be a useful addition to the surgical, endoscopic, nutritional, and drug treatments available from other members of the team.

FINDING OUT WHAT IS WRONG

The digestive system comprises approximately 26 feet of convoluted tubing—any part of which can potentially become damaged or diseased. Diagnosing the exact location and nature of a problem is therefore a daunting task, but one that has been made easier and more accurate in recent years with advances in X-ray techniques and endoscopic technology.

As well as these sophisticated investigations, procedures such as examining stool specimens and blood samples still provide doctors with vital diagnostic clues.

Medical history and examination

Initial diagnosis of digestive problems does not require sophisticated equipment or expensive tests. Skillful questioning and a thorough physical examination can provide the doctor with all the necessary information.

Symptoms in the digestive system are extremely common and often short-lived and are not usually caused by serious disease. Despite this, they can be troublesome and uncomfortable and, if persistent, often lead to a visit to the doctor; in some cases, the symptoms may be troubling enough to prompt a referral to a gastroenterologist—a hospital doctor who specializes in the digestive system.

The first priority of any doctor (whether a physician or a hospital specialist) is to take a medical history and perform a physical examination. Both of these basic procedures can provide a clear picture of the problem and give an indication of which diagnostic tests and investigations would be appropriate.

THE HISTORY

The medical history is the discussion between the patient and the doctor during which the patient "tells his or her story." Most diagnoses are made during this initial consultation—subsequent investigations and tests are mainly carried out to confirm the diagnosis.

The nature of the symptoms

Occasional episodes of the most common digestive symptoms—abdominal pain, diarrhea, constipation, nausea, and gas—are familiar to most of us and are either ignored or tolerated until they go away. Persistent or unusual symptoms that begin to affect everyday life, however, may indicate a more serious condition.

Describing the exact nature of a symptom, such as how and when it developed, can give doctors significant clues as to its cause. What brings symptoms on and what relieves them is particularly relevant. For example, pain in the stomach that starts before meals but is relieved by eating may be caused by a stomach ulcer. Unexplained weight loss should always be mentioned, especially if clothes are becoming noticeably looser—a good indication of significant weight loss.

HELP YOUR DOCTOR TO HELP YOU

Describing your bowel habits

One of the most important indicators of problems in the digestive tract is bowel habit. These symptoms can be rather embarrassing to discuss, but it is vitally important to mention such details, even if they don't seem relevant. The doctor will probably want to know the answers to the following questions.

- *How often are your bowels opened? The "normal" range can be twice a day to twice a week.*
- *Is the stool loose and watery, hard and pellet-like, or pale and foul-smelling?*
- *Is there any blood or mucus in the stool? Is there any evidence of blood on the toilet paper after wiping?*

Color, consistency, and frequency

Be prepared for questions about the regularity, consistency, and appearance of stools—the description provided may be helpful to the doctor. The trouble is that everyone is different, and it is very difficult to define what is normal—the most important question is, therefore, whether the regularity, consistency, and appearance of stools have changed. If bowel habit changes noticeably, keep a record of how often episodes of diarrhea or constipation occur and, embarrassing as it may seem, always check the appearance of the stools.

Bleeding from the rectum is particularly important, as it could be a sign of cancer. Far more often, however, rectal bleeding is caused by hemorrhoids, which, although uncomfortable, are not serious.

Constipation is not usually a symptom of serious disease, but it is important to mention if it is difficult to have a bowel movement, whether or not laxatives help, and, if so, how often they are used.

Health history

In order to get a complete health picture, the doctor will want to know about any other existing illnesses, such as diabetes, high blood pressure or heart disease. He or she may also ask about any previous surgery or blood transfusions, which may be relevant. Some symptoms, such as diarrhea or nausea, could be side effects from both over-the-counter and prescription drugs, so it is helpful to give a list of these to the doctor.

Knowledge of a family history of illness is of particular relevance, as a number of digestive disorders, for example inflammatory bowel disease and cancer of the colon, can affect more than one member of the same family.

Psychosocial history

The behavior of the digestive system can reflect a person's state of mind. Stress and anxiety, for example, can lead to symptoms such as pain and diarrhea. Stress is also known to contribute to the development of certain conditions, such as irritable bowel syndrome. Symptoms in the digestive tract are also a feature of certain psychological

Family history and the risk of colon cancer

This chart gives information on your relative lifetime risk of colon cancer, in relation to the general population, if a member of your family has been affected.

- More than 25% of patients diagnosed with colon cancer have a family history of the disease.
- People whose family members have had colon cancer face twice the risk of developing the disease than do people in the general population.
- Families who have multiple members with colon cancer can be at a tenfold risk for the disease when compared to the general population.
- If cancer develops in high-risk individuals, it frequently appears around 40–50 years of age rather than later in life.
- Of 100 colon cancer patients, 20 know of the disease among one or a few family members (not necessary having developed the disease at a young age and not necessarily on the same side of the family). About 5–10 will have a strong family history.

Taking care of business
Learning to manage time and finances effectively can indirectly help your digestive health by relieving stress and anxiety.

disorders. It is particularly important, therefore, that the doctor is aware of individual social circumstances. A high-stress job, financial struggles, or an unhappy relationship could all give rise to symptoms, as could excessive alcohol consumption and recreational drug use.

THE PHYSICAL EXAMINATION

The most important part of the doctor's physical assessment is the abdominal examination, although the doctor will also routinely take a patient's pulse and measure his or her blood pressure.

Look, feel, and listen

The doctor will ask the patient to undress down to underwear and will maintain privacy throughout the procedure by using a blanket. There are three parts to the examination: visually assessing the abdomen (inspection), feeling for abnormalities (palpation), and listening to the abdomen through the stethoscope (auscultation).

- Auscultation The doctor will listen to the abdomen with a stethoscope to listen for "bowel sounds"—the tinkling noises made by the normal working of the intestines—which may be unusually loud or quiet, or even absent, in certain conditions.
- Palpation The doctor will gently feel the abdomen for tender areas, lumps or bumps, or enlarged organs. The doctor may then percuss (tap with the fingers) over particular areas. Different sounds indicate hollow or solid tissue, and it is possible to tell whether an organ, for example the liver, is enlarged.
- Inspection Often, simply looking at the abdomen can provide a great deal of information. Some conditions, for example bowel obstruction, cause bloating or distension and if severe, the abdomen will look swollen and tense. The doctor will also be on the lookout for any visible bruises, blemishes, or operation scars.

Examining the abdomen
a Tensing of the abdominal muscles in response to palpation is common in patients with pain or tenderness. It is known as guarding and is a symptom in itself.
b Bowel sounds should be present, but not too loud. A stethoscope can also be used to check the liver margins—scratching over the liver produces a soft sound.

Checking the rectum

If bowel symptoms suggest an abnormality in the rectum, and in particular if there is any sign of blood, a rectal examination may be necessary. The examination is quick and can be done in the doctor's office, with the patient lying on the left side with his or her knees drawn up. The doctor gently inserts a gloved and lubricated finger through the anus into the rectum to feel for any tender areas or lumps. Although the procedure may be uncomfortable, it is not usually painful. Sometimes the doctor will perform a sigmoidoscopy or proctoscopy, where a small, telescope-like device is inserted through the anus to allow the doctor to check the rectum visually.

Investigative tests

A doctor usually makes a tentative diagnosis after the initial examination. Further tests are then used to verify the hypothesis and to help eliminate other possible causes of illness.

Digestive symptoms, although uncomfortable, do not necessarily require hospital admission. Many of these tests therefore, which are mostly straightforward and do not need complex equipment, can be carried out by a person's own physician or in an outpatient clinic.

BLOOD TESTS

Blood analysis provides a wealth of valuable information and is simple and quick to perform. Using a needle and syringe, enough blood for several samples can be taken at one time from a vein in the arm and sent to the laboratory—each in a different colored tube for ease of identification—to be tested.

When blood is spun at very high speed in a centrifuge, the blood cells separate from the plasma—the straw-colored fluid in whereas they float. Some blood tests look only at the blood cells, whereas others measure substances that are dissolved in the plasma.

Using a centrifuge
Different elements of blood have different weights. Spinning blood at high speeds in a centrifuge separates the components so that they can be investigated independently.

Counting cells

There are three types of blood cell, all of which carry out different functions: red cells, white cells, and platelets. Red cells are responsible for carrying oxygen to the body's tissues. The white cells, of which there are several different types, play an important role in defending the body from invaders, such as infectious organisms. Platelets are a group of cells that help blood to clot if necessary. A severe reduction in the level of any of these blood cells could have serious consequences.

From only a small sample of blood, a machine can provide a count of each type of cell. This information can give doctors clues about a patient's condition. For example, a low red cell count indicates anemia and may suggest that there is bleeding in the digestive tract. The white blood cell count may be elevated in someone who has an infection or inflamed areas in the digestive tract. Platelets also increase in number if there is inflammation or bleeding in the bowel.

A test called the erythrocyte sedimentation rate (ESR) does not actually count the cells but measures how long it takes red blood cells (erythrocytes) to settle in a sample of blood. A raised rate of settling is a general indicator of inflammation or infection.

If I have a lot of blood taken for a test, will I feel ill? How does it get replaced?

You would need to have a great deal of blood taken before you felt unwell: Blood donors lose slightly more than a pint each time—much more than in a blood test—generally with no adverse effects. In addition, your body won't really miss the blood. Blood cells are produced continuously in the bone marrow. If there is a significant drop in your blood volume, a feedback mechanism swings into action and the marrow speeds up production of blood cells.

ASK THE EXPERT

Milestones IN MEDICINE

The first person to see bacteria under the microscope was a Dutch shopkeeper, Anton van Leeuwenhoek, who made lenses for a hobby and then examined very small things. On a normal working day in 1674, Anton took a scraping from between his teeth and observed it through one of his lenses: "I then saw, with great wonder . . . many very little animalcules, very prettily a-moving". Presenting his findings to Britain's prestigious Royal Society, van Leeuwenhoek initially met with disbelief and hostility, but later became a member of the organization. As his fame spread, the former shopkeeper, now a civil servant, received visits from the rich and famous, among them the Queen of England and the Russian Czar.

Measuring substances in blood

In addition to looking at blood cells, doctors may also need to measure various chemicals that are dissolved in the plasma. In a patient with suspected pancreatitis, for example, doctors may take a blood test to measure a substance called amylase, levels of which shoot up when the pancreas is inflamed. Often, a person with a digestive tract disorder is poorly nourished because of absorption problems or a loss of appetite. Measuring levels of proteins such as albumin in the plasma can confirm that a person is not absorbing nutrients because of GI disease or is not making sufficient proteins in the body.

In a condition called celiac disease, the body produces antibodies as part of a sensitivity to gluten, a protein found in wheat, rye, and barley. Using sophisticated equipment, experts can detect the presence of these abnormal antibodies in the blood. Physicians sometimes use a blood test to detect antibodies against the bacterium *Helicobacter pylori*, which is known to contribute to the development of stomach ulcers. This confirms whether a person has been infected with the organism during his or her lifetime. It is not, however, useful in assessing the effectiveness of treatment, because even when the infection clears, the antibodies remain in the bloodstream.

Checking the liver

Some tests assess how well the liver is working and whether it is damaged by measuring substances that are either produced or broken down by the liver. Certain enzymes are excreted by damaged liver cells and these can also be measured. Testing the blood for levels of clotting factors can be an indicator of liver function. Many of the factors that produce normal blood clotting are made in the liver, and in some people with liver problems, levels of these may be abnormal.

Measuring minerals

Mineral salts are dissolved in the plasma and are necessary for the normal functioning of the body. Looking at the levels of some of these minerals, such as potassium, calcium, and magnesium, in the blood can be very useful in the assessment of people with diseases of the digestive tract. Low levels of potassium are frequently found in people with long-term diarrhea, wheras people with conditions such as Crohn's disease and celiac disease, who are not absorbing food properly, tend to have low levels of calcium or magnesium.

STOOL TESTS

Apart from a simple visual inspection of the stool in the outpatient clinic or in the hospitl, which may reveal blood, mucus, or an abnormal color and consistency, specimens are sometimes sent for laboratory testing. A patient is given a clean container to take home or into the bathroom, where a small amount can be collected in privacy. The sample is then sent to the laboratory for testing. There are three main factors that doctors will be curious about: how much stool is

Coeliac disease
This intestinal wall is flat, pitted and without villi. Compare this with the healthy one on *page 26.*

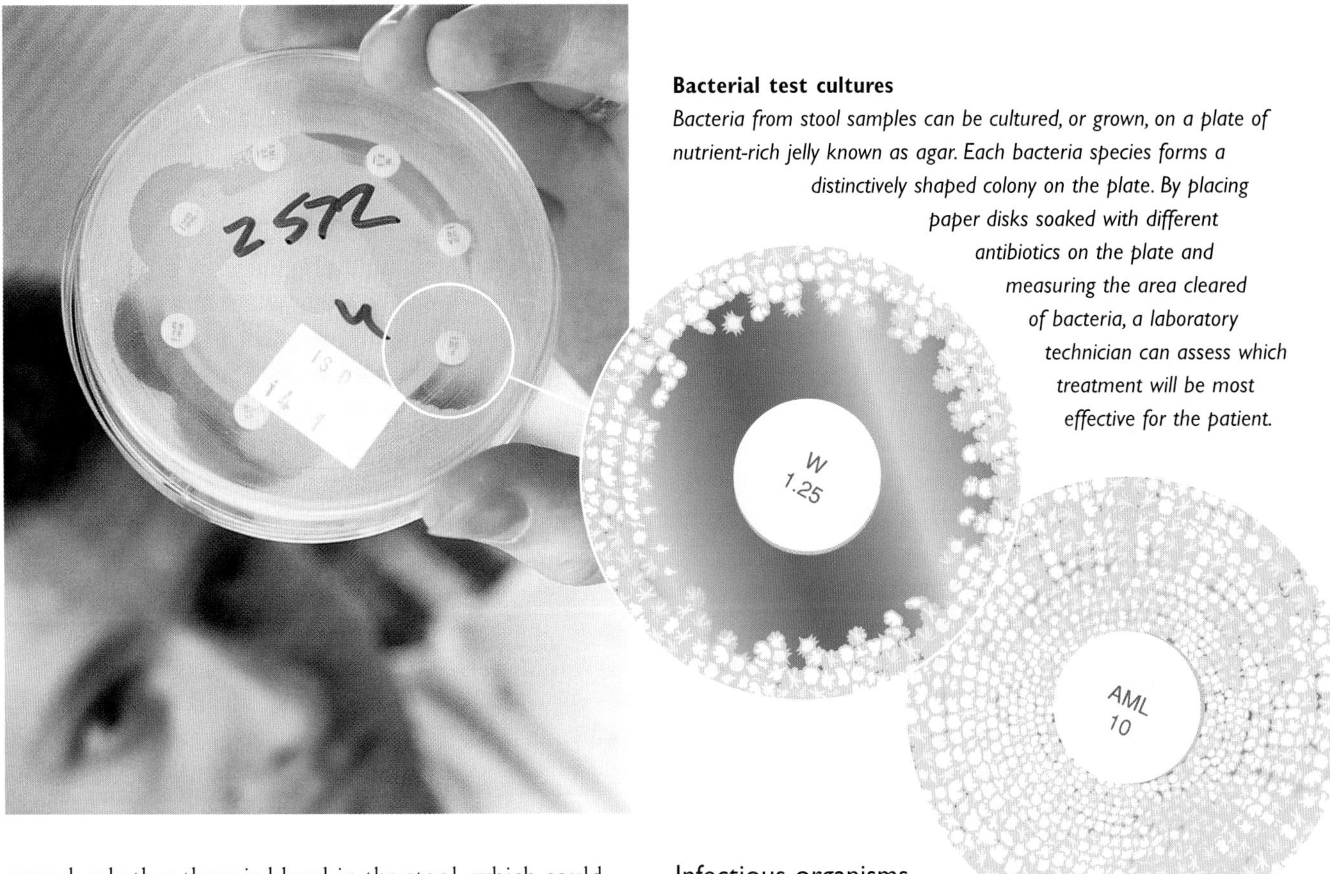

Bacterial test cultures

Bacteria from stool samples can be cultured, or grown, on a plate of nutrient-rich jelly known as agar. Each bacteria species forms a distinctively shaped colony on the plate. By placing paper disks soaked with different antibiotics on the plate and measuring the area cleared of bacteria, a laboratory technician can assess which treatment will be most effective for the patient.

passed; whether there is blood in the stool, which could indicate a damaged GI system; and whether there are any abnormal microorganisms, such as viruses or parasites, that might be causing infection.

Looking for blood

Sometimes, the presence of blood is clearly visible to the naked eye. If the bleeding point is in the rectum, streaks of bright red blood will coat the stool. Blood from higher up in the system will cause the stools to look very dark and tarry, even black.

Sometimes, though, doctors may suspect that a digestive disorder is causing very slight bleeding that is not visible—this is technically known as occult ("hidden") bleeding. In these cases, fecal occult blood testing (FOBT) may be carried out. A tiny sample of feces is placed on an absorbent card and a chemical added. The presence of even a microscopic amount of blood causes the card to change color. The test is usually repeated three times over a few days because blood may not be present in every sample of feces.

The major drawback of this test is that it frequently produces a positive result even when there is no bleeding from the GI system. In fact, the test is so sensitive that even bleeding gums can produce a positive result.

Infectious organisms

There are some bacteria that are known to cause diarrheal illness if they are present in the GI system. These bacteria include certain types of *Escherichia coli, Campylobacter, Salmonella*, and *Shigella*. If the doctor suspects that a patient has been infected with one of these bacteria, a stool culture test may be arranged. For the test, the patient provides a small sample of stool, which is sent to the laboratory. It is then cultured on a Petri dish of agar jelly in warm conditions to encourage any bacteria to grow. After a period of time, the specimen is treated with staining agents and examined under the microscope—bacteria can be identified by their size and shape, and the way they react with the stains. If a pathogen is identified, it is possible to measure its sensitivity to a battery of antibiotics so that the appropriate drug can be prescribed.

BREATH TESTS

Currently, the only breath test routinely used in gastroenterology is the test for *Helicobacter pylori* bacteria, the organisms strongly associated with stomach ulcers. This is a simple test that can be done in a doctor's office or a hospital clinic.

Helicobacter pylori
This species of bacteria has characteristic flagellae—whiplike structures—at one end.

Saving your breath

To begin, the person swallows a small capsule containing a chemical called urea, which is labeled with a non-radioactive marker. A breath sample is taken to provide a control reading of carbon dioxide levels. After 45 minutes another sample of breath is collected, and both breath specimens are analyzed for the marker substance. If a person's stomach is infected with *Helicobacter pylori*, the organism metabolizes the urea in the capsule. This process produces carbon dioxide, which is excreted in the breath and can be measured because of the marker. This test is ideal for detecting the presence of *Helicobacter pylori*, as well as for assessing whether the bacteria have successfully been eradicated by a course of treatment.

About half the world's population is infected with* Helicobacter pylori, *but only a small proportion of people develop peptic ulcers.

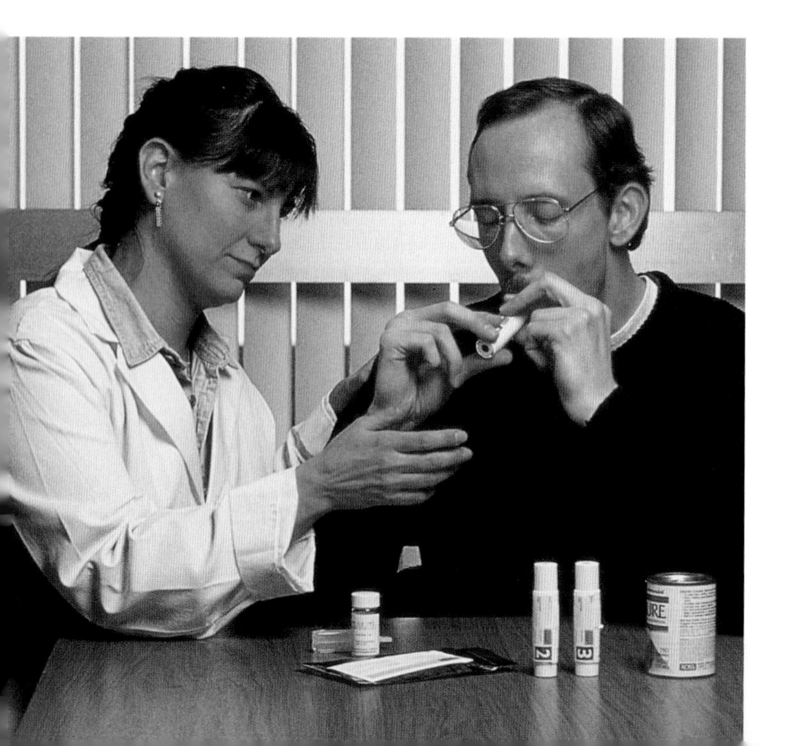

pH MONITORING AND ESOPHAGEAL MANOMETRY

These specialized tests are used in people with severe heartburn who are not responding to normal treatment, or who are having difficulty swallowing that is not caused by esophageal narrowing. They are used to measure the pressure and the levels of acid in the esophagus over a 24-hour period.

Acid test

In pH monitoring, a doctor or nurse passes a small pH-sensitive probe through the nose and down into the esophagus where it is left in position. The person is then sent home, and over a 24-hour period the amount of acid reflux is recorded by a small device worn on the body. The person keeps a diary so that a correlation between heartburn and other reflux symptoms can be made.

Under pressure

Esophageal manometry measures the pressure in the esophagus during the process of swallowing and is useful in diagnosing conditions that affect motility, such as achalasia. During the normal swallowing process, a morsel of food is propelled down the esophagus toward the stomach by muscular contractions called peristaltic waves. The muscle at the lower end of the esophagus then relaxes to allow food to enter the stomach. If the muscles of the esophagus are not contracting and relaxing properly, unusual changes in pressure result. Esophageal manometry seeks to measure these changes.

A doctor or nurse passes a small catheter through one nostril until it is positioned in the esophagus, and the person is asked to swallow. Changes in pressure in the esophagus are recorded by a gauge attached to the catheter. At the end of the procedure, the catheter is removed and the patient goes home.

Scenting danger
Here a nurse assists a man as he breathes into a sample tube, in a test designed to detect the presence of Helicobacter pylori *bacteria. Levels of carbon dioxide in the sample are measured.*

Imaging the digestive system

Many of the soft tissue structures of the abdomen cannot be seen with traditional X-ray techniques. However, in the past two decades there have been exciting developments in imaging technology, and the field of digestive medicine has particularly benefited.

It is now possible to obtain detailed, real-time three-dimensional pictures of the organs of the digestive system. This makes diagnosis quicker, easier, and more accurate, and helps to target and assess treatment options more effectively than ever before.

ULTRASOUND SCANNING

Ultrasound scanning is a simple and safe method of imaging abdominal organs without the use of radiation. Most people will be familiar with ultrasound; because of its safety, it is routinely used to examine fetuses in pregnant women.

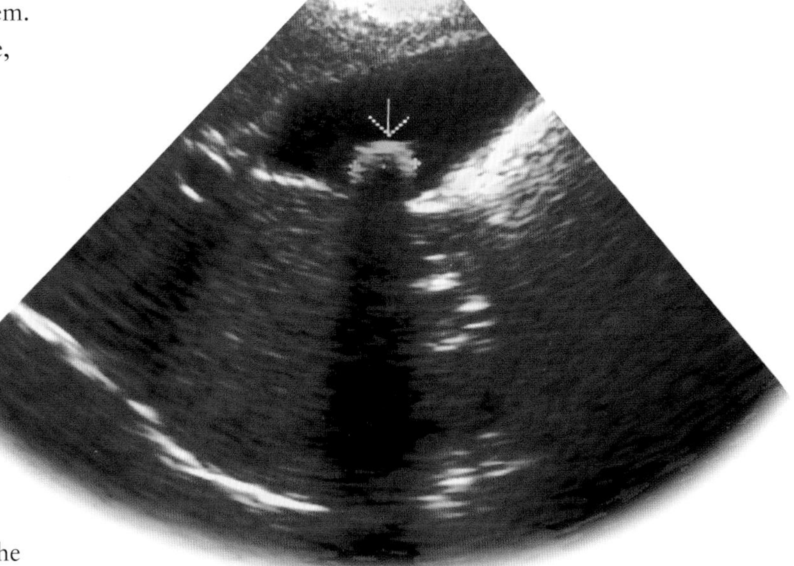

Echoes of stone
This colored ultrasound scan shows a gallstone (see arrow), an accretion of fats and minerals, in a man's gallbladder. The stone could become stuck in the bile duct, causing excruciating pain.

What is it used for?

Ultrasound scanning has limited ability to image hollow organs, such as the intestines, but it is usually the first choice of doctors for an initial look at the pancreas, liver, bile ducts, and gallbladder. Abnormalities such as gallstones, tumors, and abscesses are easily located, and doctors can measure the size of the organs.

If an abnormality is noted on an ultrasound scan, for example a possible tumor, doctors can use the technique to help guide a small needle to take a sample for further analysis. Ultrasound can also be used to guide the draining of abscesses or fluid from within the abdomen.

ON THE CUTTING EDGE

Endoscopic ultrasound

An ultrasound probe can be attached to an endoscope—a flexible viewing instrument—and passed into and down the digestive tract. This "internal ultrasound" can be used to detect the presence of a tumor and measure the thickness of the esophagus and stomach walls; it is one of the most accurate ways of diagnosing cancer of the pancreas. However, this technique is very difficult to master and not yet widely available.

How does it work?

Ultrasound works by firing rapid high-frequency sound waves—too high-pitched for the human ear to hear—at the body's tissues, measuring the echoes as they bounce back and converting them into visual images that are shown on a screen. Different types of body tissue reflect different echoes and, using modern equipment, a skilled operator can obtain extremely detailed images of the internal organs. Also, because images are produced in "real time"—as they happen—movement can be detected.

An abdominal ultrasound is carried out with the patient lying almost flat on a bed or gurney. The ultrasonographer or doctor covers the skin with conducting jelly and then moves a probe, known as a transducer, gently over the area being scanned, observing the image on the screen and taking "still" photographs when necessary. Usually, the patient can watch the screen during the scan. The procedure is painless and completely safe, and usually takes between 10 and 30 minutes.

COMPUTED TOMOGRAPHY (CT) SCANNING

CT scanning is a sophisticated X-ray technique introduced in the early 1970s—about 80 years after X-rays were first used to look inside the body. The technique is generally carried out only after some other investigation has raised the possibility that something may be amiss.

What is it used for?

CT scanning gives excellent images of the digestive tract and is particularly useful for the detection of tumors and abscesses. Like ultrasound, CT scanning can be used as image guidance when an abnormal area is identified for the doctor to take biopsies using specialized needles.

How does it work?

The technique uses X-rays to produce a series of cross-sectional slices of the body using computerized reconstruction. A rotating X-ray source sends multiple narrow beams through the body. As they exit, they are detected and analysed by equipment that is sensitive to the variations in tissue density and that reproduces these variations on the resulting image in different shades of gray. The many cross-sections are analysed by computer to form a "map" of the body area.

Abdominal CT scans are often done to look at the organs of the digestive system. Because many of these are hollow, X-rays travel straight through them and the resulting images are unclear. To overcome this difficulty, on arrival at the X-ray department, the patient is usually given a drink called gastrografin. This is a contrast medium that absorbs X-rays, outlining the stomach and small intestine so that they are easier to identify on the CT images. Sometimes, as well as drinking gastrografin, contrast is enhanced by injecting a contrast medium into the bloodstream during the procedure.

Are there any adverse effects?

The main concern about CT scanning is the exposure to X-rays. Because a series of images are required, it can add up to quite a high dose of radiation, even though the dose for an individual image is small. The risk, however, is minimal and exposure is carefully monitored (see box). Additional concerns include the use of the contrast dye, especially in patients with allergies or kidney problems.

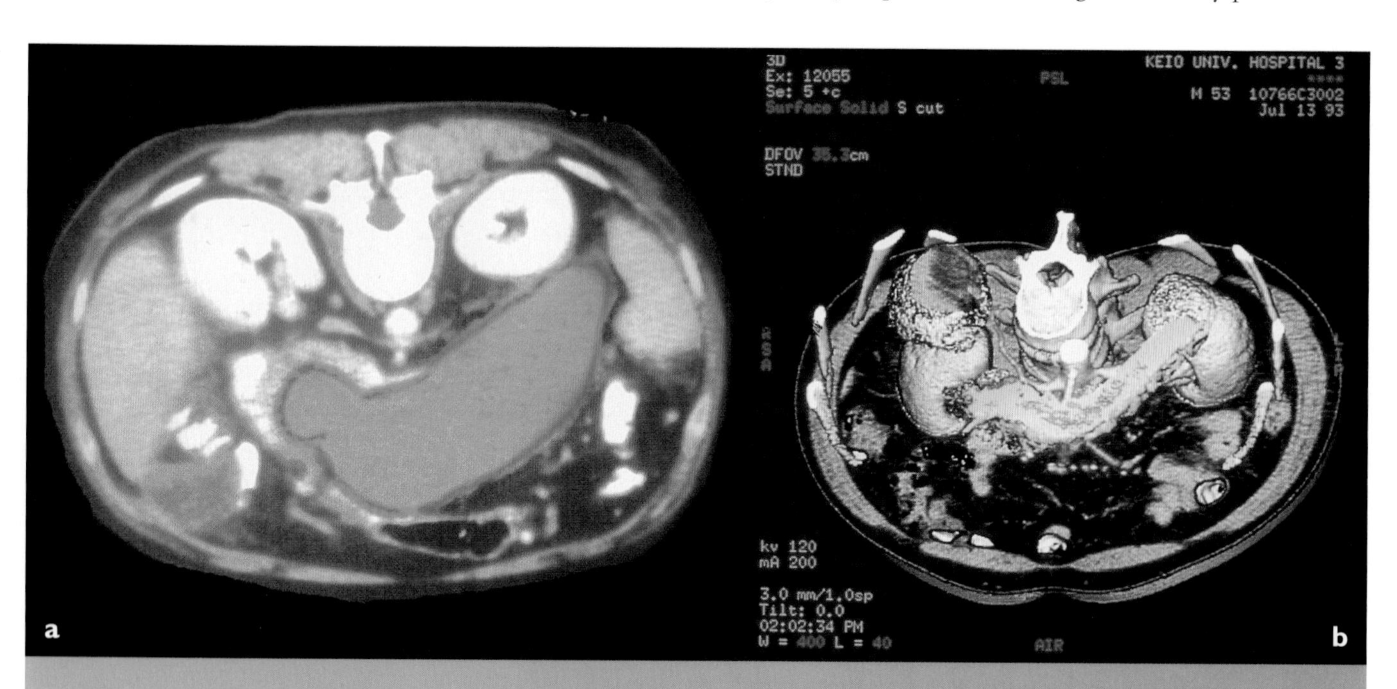

CT scanning: slices of life

a This scan shows a horizontal "slice" through the torso of a pancreatitis sufferer. The spinal cord is visible at the top, with the kidneys on either side. The blue mass is an enlarged pancreas.

b This advanced 3-D CT image shows a healthy torso. The spine (beige), kidneys (blue–gray), gallbladder (green–gray), and a normal pancreas (yellow) are visible, ringed by the ribs.

Is radiation-based imaging safe?

Everyone is exposed to radiation in daily life, from cosmic rays from outer space to radioactive rocks in the earth. The important safety question about radiation exposure is therefore not "whether" but "how much." The radiation exposure a patient receives from one session of CT scanning is not dangerous, especially because the radiation doses are kept as low as possible. However, special measures should be taken to protect reproductive organs, and pregnant women should avoid X-rays altogether, as there is a very small risk that high doses of radiation may cause birth defects.

ASK THE EXPERT

MAGNETIC RESONANCE IMAGING

Magnetic resonance imaging (MRI) is a relatively new imaging technique that is rapidly becoming more widely used, where available, because of its ability to produce amazingly detailed pictures of the body's interior.

What is it used for?

In gastroenterology, doctors use MRI mainly to look at the bile ducts, gallbladder, and liver. Abnormalities such as gallstones, cysts, abscesses, and tumors are very clearly defined on MRI. Because of the fine detail, the technique is also helpful in detecting abscesses around the anus.

How does it work?

MRI uses a powerful magnet and radio waves to create 3-D images on a computer screen. Human tissue is mainly composed of water, which contains hydrogen atoms. When a patient is exposed to a huge circular magnet, the pattern of the hydrogen atoms in the body is disrupted so that rather than existing randomly, they lie in rows pointing in the same direction.

The appropriate part of the body is then bombarded with radio pulses, which knock the atoms temporarily out of alignment. As they fall back into place, the nuclei in the atoms emit signals—the so-called "resonance"—that are picked up and analyzed by a computer, according to their strength and source, to form clear images. Sometimes a contrast medium is used to enhance the definition of the pictures obtained.

MRI generates images on a computer using a magnetic field that is between 10,000 and 30,000 times more powerful than the earth's magnetic field.

What are the adverse effects?

Although the process of MRI sounds extraordinary, patients undergoing the procedure do not feel a thing. As far as the experts know, MRI is an extremely safe procedure, because no radiation is involved. However, the magnets can interfere with pacemakers, hearing implants, and other artificial devices. The experience itself can be uncomfortable because the subject spends up to 30 minutes in a narrow tube while the image is obtained, and patients who suffer from claustrophobia may need a sedative beforehand to calm them.

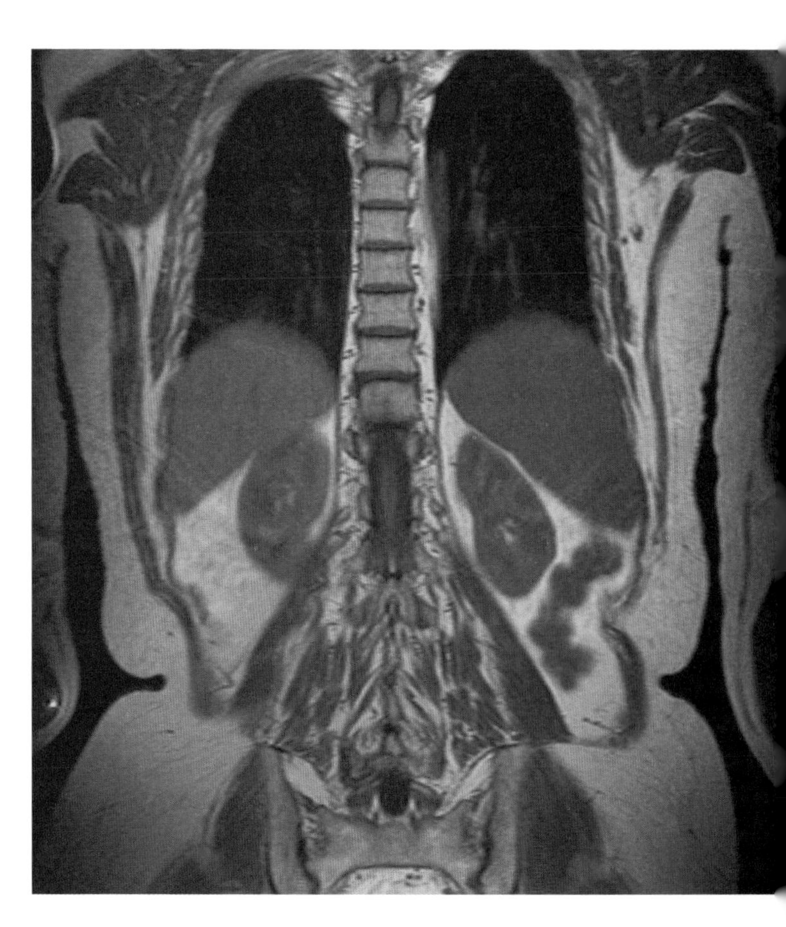

Sliced in half

This MRI shows a section of a torso, seen from the back. The spine runs down the center, flanked by the lungs at the top, and two lobes of the liver, two kidneys, and a short length of intestine on the right.

Barium studies

For more than a hundred years, barium studies have been the primary technique for imaging the gastrointestinal tract. They have the advantages of being cheap, quick, and non-invasive. A century of development has led to better media, new methods, and advanced X-ray equipment, which combine to provide extremely high-quality images.

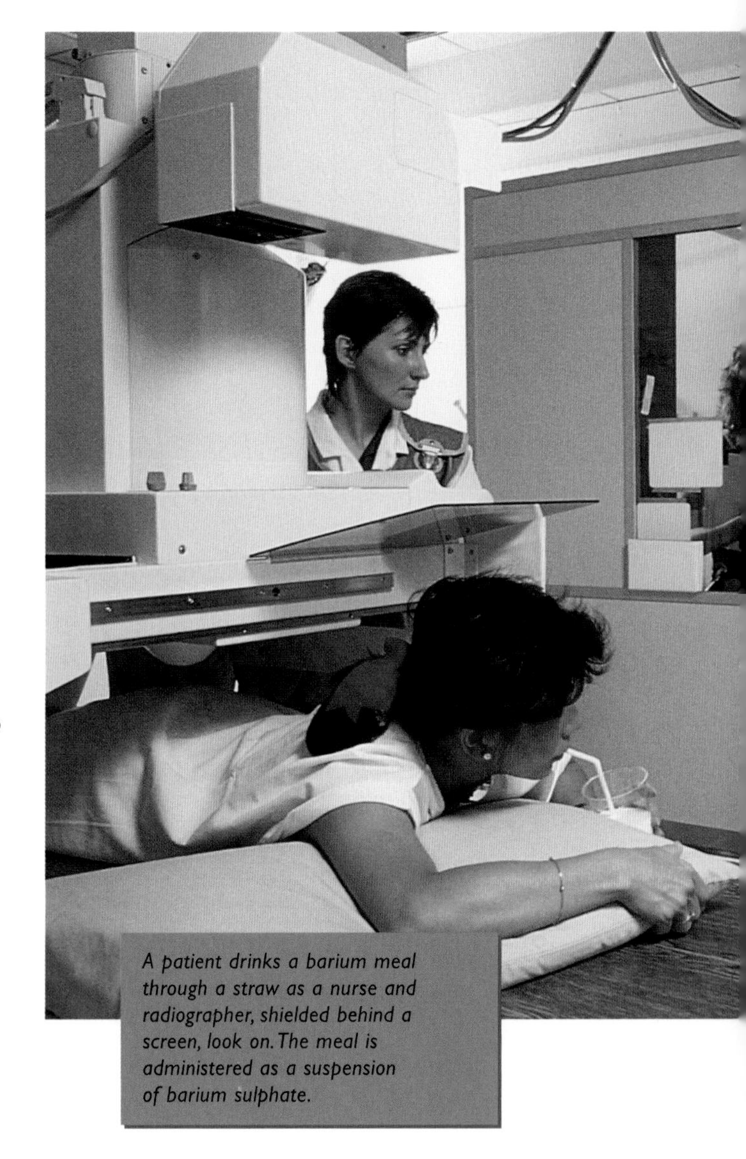

A patient drinks a barium meal through a straw as a nurse and radiographer, shielded behind a screen, look on. The meal is administered as a suspension of barium sulphate.

BARIUM CONTRAST STUDIES

With the introduction of X-rays in 1895, doctors were finally able to look inside the human body without having to cut a patient open. However, the images obtained depended on the ability of the various body tissues to absorb X-rays. Those that absorb X-rays well, such as bone, are reproduced beautifully on the radiographs. But the amazing new technology appeared to have limited use in visualizing organs such as the stomach and intestines, which did not show up at all well on the X-rays because they are hollow.

Within a year, however, the principle of a "contrast medium" was developed. In 1896, it was discovered that when cavities are filled with barium, a substance that readily absorbs X-rays, a clear outline can be seen. To give an even clearer picture, air can be introduced into the cavity once the barium is in place, a technique known as a double-contrast X-ray.

Walter Cannon was still a first-year medical student when, in 1896, he devised the use of barium as a contrast medium to outline the upper digestive tract.

What are they used for?

There are four main types of barium studies, each of which examines a different section of the digestive tract.

- Barium swallow This is mainly carried out to investigate patients who have difficulty swallowing. The test shows whether the process of swallowing is normal and indicates any abnormality of the esophagus, such as a narrowing (known as a stricture). A barium swallow may also be used as a test for gastroesophageal reflux, a condition in which some of the contents of the stomach pass back up into the esophagus.
- Upper gastrointestinal (GI) barium study This type of study is effective in visualizing abnormalities in the stomach and duodenum, for example a hiatal hernia or cancer. Before the study, the patient drinks a barium meal.
- Small bowel follow-through The aim of this type of study is to follow the barium beyond the lower part of the stomach into the small intestine to look for inflammatory conditions such as Crohn's disease.
- Barium enema This procedure visualizes all of the large bowel and is carried out if doctors suspect conditions such as ulcerative colitis or colon cancer. The technique does not always reliably image the rectum.

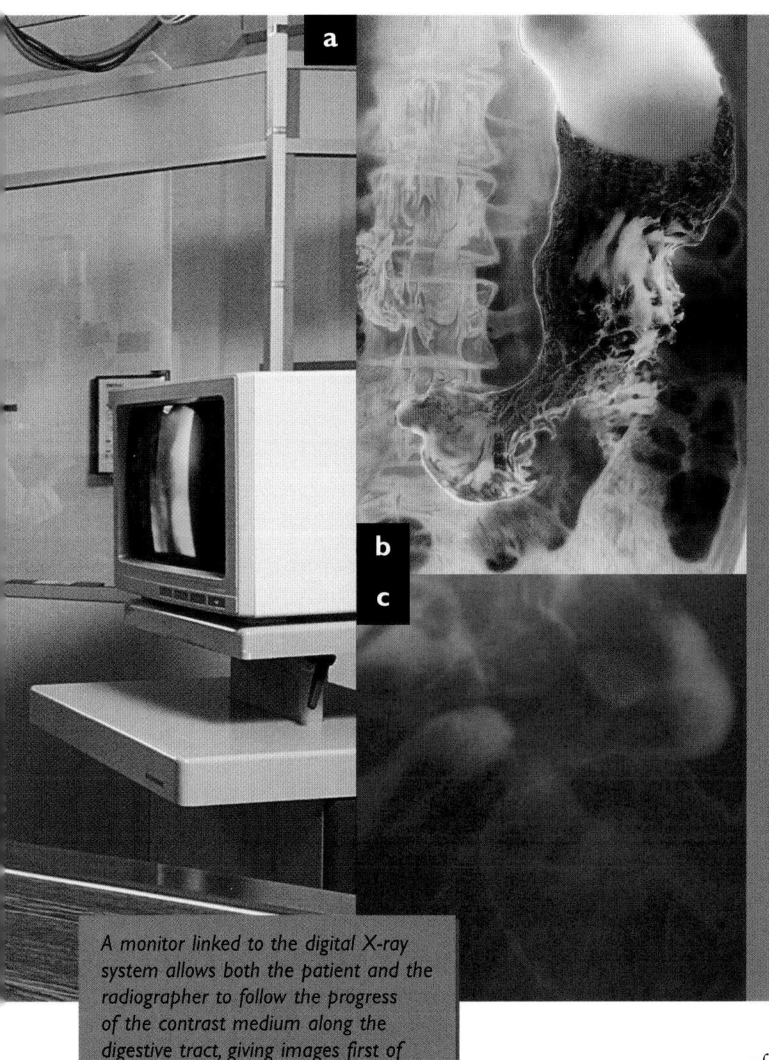

A monitor linked to the digital X-ray system allows both the patient and the radiographer to follow the progress of the contrast medium along the digestive tract, giving images first of the stomach and then of the intestine.

Upper GI studies

a If doctors suspect a problem with the stomach or small intestine, such as an ulcer or some form of inflammatory condition, an upper GI study can provide a good picture of the stomach and intestinal lining, highlighting damaged or deformed regions of the mucus membrane. Barium meals are usually followed by eating gas-producing effervescent granules, so that the digestive tract is filled with air while the barium forms a thin coating on the inside surfaces. This shows up features of the GI surface that could not otherwise be seen.

b This false-color upper GI X-ray reveals an advanced cancer that has spread across almost the whole bottom half of the patient's stomach. The uniformly purple area at the top is the healthy part of the stomach, filled with barium. In stark contrast, the lower parts of the stomach appear strangulated and wasted where the cancer has taken over. At the left, the patient's spinal column is visible, leading down to the pelvic bones at the bottom of the picture.

c This false-color double-contrast X-ray was taken after a carbonated barium meal was administered. The image shows the patient's duodenum, revealing the formation of a peptic ulcer, highlighted as a pink oval in the duodenal cap.

How do they work?

Barium sulphate is a chalk-like liquid that is either swallowed so that it coats the upper digestive tract or given as an enema into the large bowel. Once the barium is in the GI system, X-rays are transmitted through the body onto a photographic plate. As the rays pass through, the contrast medium slows their progress and shows up as a white area on the film. At the same time, images are transmitted to a monitor for real-time viewing.

Barium swallow, upper GI, and small bowel follow-through

For most barium studies, the patient is given a set of instructions on how to prepare for the test and where it will take place—which is usually in the X-ray department of a hospital. A period of fasting is always necessary for a few hours beforehand to ensure that the stomach or intestines are completely empty. Once they're underway, barium studies usually take less than an hour.

In a barium swallow procedure, the patient drinks some barium, initially while sitting up and then when lying down—changing position helps to move the barium through the GI system. X-rays are taken as it passes down the esophagus and into the stomach. An upper GI is similar, but only a small amount of barium is drunk, together with some effervescent granules or tablets, which produce gas to allow double-contrast X-rays to be taken.

For a small bowel follow-through, a slightly different technique is used. The patient may be asked to lie in different positions on the X-ray table to help move the barium through the digestive tract. Some hospitals have specially designed tables that rotate and tip during the procedure, which is important for immobilized patients.

HELP YOUR DOCTOR TO HELP YOU

Preparing for a barium enema

In order to get a good outline on the X-rays taken during a barium enema, the large intestine should be completely clear. The procedure used to achieve this can be unpleasant in that it produces frequent watery stools, but it is vital that you stick to the preparation schedule so the investigation goes smoothly. Different hospitals vary slightly in their procedures.

- *On the day the barium enema takes place you should not eat any solid food—just drink plenty of clear fluids.*
- *Drink the oral laxative supplied to you. This is usually in the form of salts, which are dissolved in a great deal of water.*
- *Stay close to a toilet, because the laxatives will cause frequent episodes of watery diarrhea.*

Barium enema

Looking at the large intestine is a more complicated procedure than the other studies—it is a long and convoluted organ, which has to be empty in order to get a good view. Consequently, a bowel preparation procedure is necessary (see left).

The investigation involves placing a small tube into the rectum through the anus and injecting barium liquid and air to outline the entire colon so that X-rays can be taken.

Are there any adverse effects?

The main side effect of barium investigations is a feeling of bloating and flatulence. In addition, because of the chalky nature of the barium, some people suffer constipation for a day or two after the test.

Because X-rays are a form of radiation, there is a certain amount of exposure. However, this dose is not high enough to cause damage, and exposure is closely monitored by radiologists and radiographers.

HAVING A BARIUM ENEMA

Three days before the procedure the doctor gave me a leaflet about following a low-fiber diet. She also gave me some strong laxatives to empty my bowel the night before the test.

On the morning of the test, I had to skip breakfast—having food in my digestive system might ruin the picture—but I did have a large glass of fruit juice, which kept me going. At the hospital I changed into a gown (completely open in the back, of course).

The doctor asked me to lie on a special table and warned me that the next part of the procedure would feel a bit uncomfortable. First, I had an injection to relax the muscle of the bowel wall, then a special tube was inserted into my rectum. This was used to run a special dye that shows up on X-rays into my intestines, which were also pumped up with air. Again, this was uncomfortable, as it caused some sharp, colicky pain in my abdomen.

Once the air and dye mixture was right, the staff retreated behind a screen and took lots of X-rays, moving the table (and me) into a different position for each one. It only took about ten minutes.

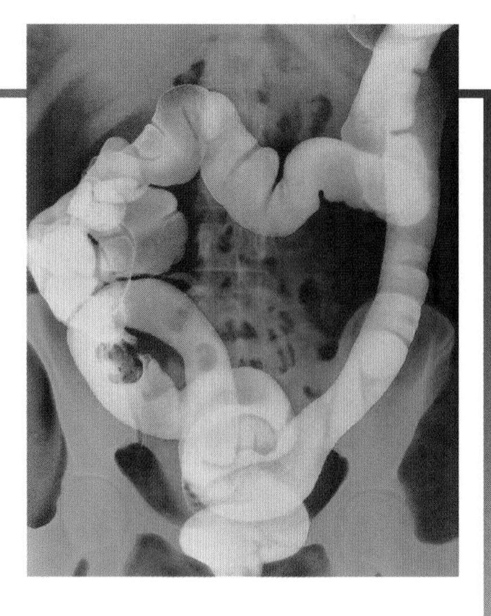

After the X-rays were finished, I was taken into a side room with a toilet to recover. I passed a lot of dye and air into the toilet right away, but as the muscle relaxant wore off, I got a bit of discomfort—along with embarrassing noises—from being bloated with gas.

Looking inside the digestive tract

The development of endoscopy over the past 30 years has revolutionized the medical specialty of gastroenterology. Video-endoscopes can now produce high-quality images and allow direct imaging of both the upper and the lower digestive tract.

Endoscopy literally means "looking inside," and endoscopic procedures of the digestive tract do just that. For nearly 40 years, these techniques have been used mainly for diagnosis—to check for abnormalities and to take samples for further examination—but they are now increasingly used to carry out treatments. Although in some procedures light sedation may be necessary, there is no need for surgery, a general anaesthetic, or surgical incisions. Because of this, endoscopic procedures can often be carried out on an outpatient basis.

Flexible endoscopes

These are narrow, bendy tubes that can be manipulated around tricky corners. They consist of bundles of optical fibers, made from either glass or plastic, that transmit light signals using reflection. Some of the fibers carry light to the GI system to illuminate it and others transmit the image back. The doctor then views the images through an eyepiece or on a monitor. Over the last few years, a new type of endoscope—the video-endoscope—has effectively supplanted the older model. This endoscope has a miniature camera at its tip, which relays high-quality video images to a monitor screen.

Both types of endoscope usually have several channels—one for viewing, one for lights, and one for sucking out fluids and blowing air. This channel can also carry instruments, such as biopsy forceps.

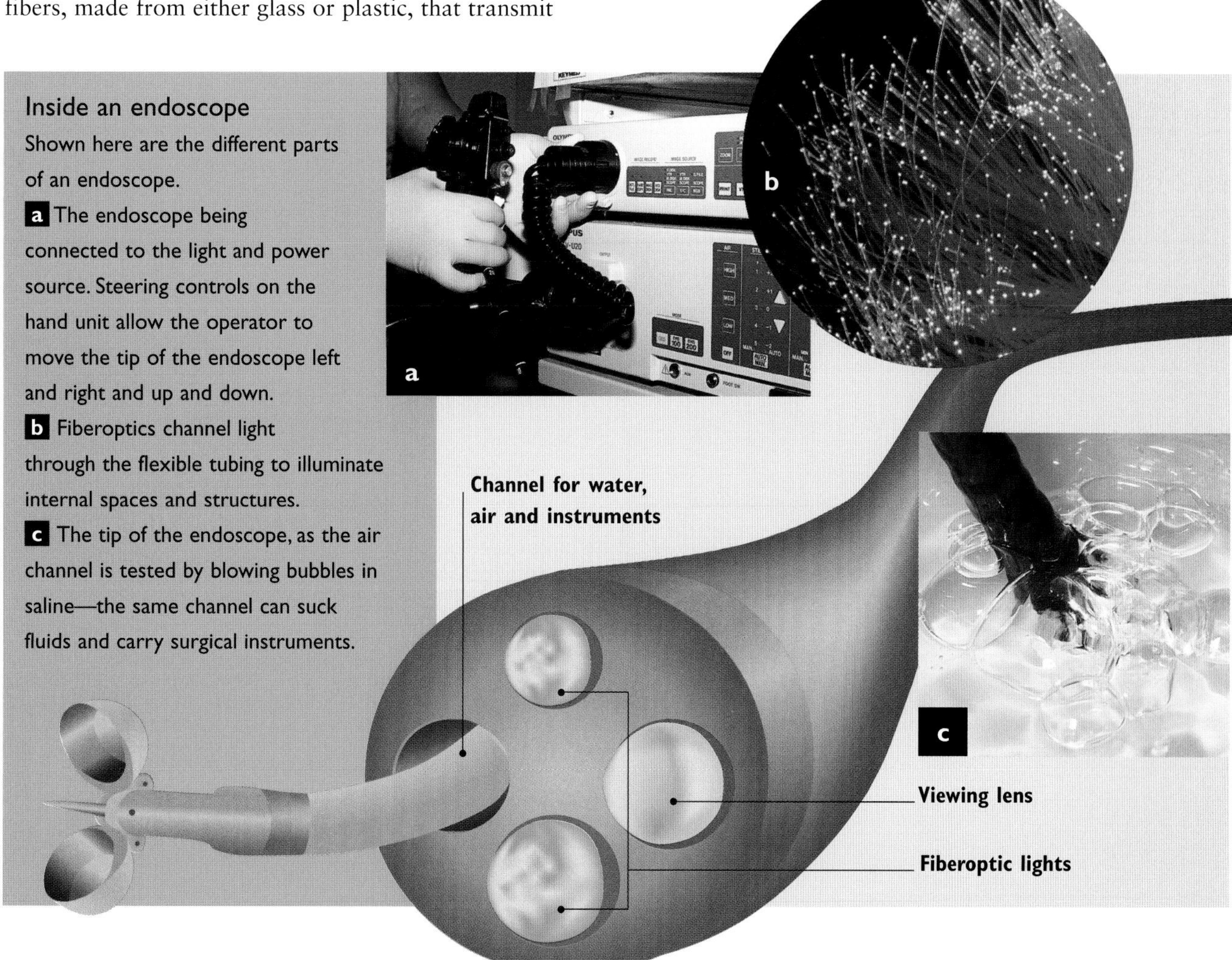

Inside an endoscope

Shown here are the different parts of an endoscope.

a The endoscope being connected to the light and power source. Steering controls on the hand unit allow the operator to move the tip of the endoscope left and right and up and down.

b Fiberoptics channel light through the flexible tubing to illuminate internal spaces and structures.

c The tip of the endoscope, as the air channel is tested by blowing bubbles in saline—the same channel can suck fluids and carry surgical instruments.

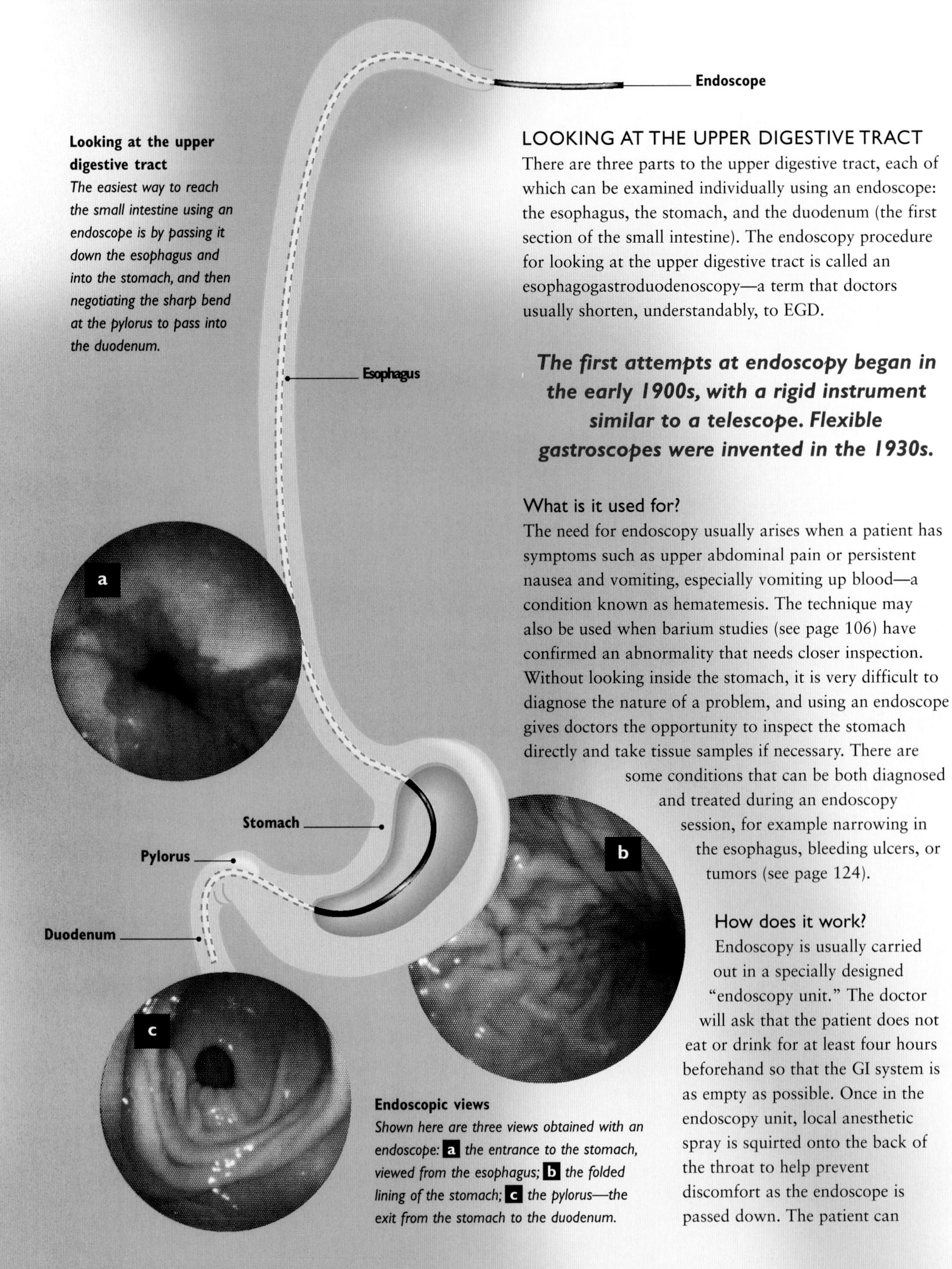

Looking at the upper digestive tract
The easiest way to reach the small intestine using an endoscope is by passing it down the esophagus and into the stomach, and then negotiating the sharp bend at the pylorus to pass into the duodenum.

Endoscopic views
Shown here are three views obtained with an endoscope: **a** *the entrance to the stomach, viewed from the esophagus;* **b** *the folded lining of the stomach;* **c** *the pylorus—the exit from the stomach to the duodenum.*

LOOKING AT THE UPPER DIGESTIVE TRACT

There are three parts to the upper digestive tract, each of which can be examined individually using an endoscope: the esophagus, the stomach, and the duodenum (the first section of the small intestine). The endoscopy procedure for looking at the upper digestive tract is called an esophagogastroduodenoscopy—a term that doctors usually shorten, understandably, to EGD.

The first attempts at endoscopy began in the early 1900s, with a rigid instrument similar to a telescope. Flexible gastroscopes were invented in the 1930s.

What is it used for?

The need for endoscopy usually arises when a patient has symptoms such as upper abdominal pain or persistent nausea and vomiting, especially vomiting up blood—a condition known as hematemesis. The technique may also be used when barium studies (see page 106) have confirmed an abnormality that needs closer inspection. Without looking inside the stomach, it is very difficult to diagnose the nature of a problem, and using an endoscope gives doctors the opportunity to inspect the stomach directly and take tissue samples if necessary. There are some conditions that can be both diagnosed and treated during an endoscopy session, for example narrowing in the esophagus, bleeding ulcers, or tumors (see page 124).

How does it work?

Endoscopy is usually carried out in a specially designed "endoscopy unit." The doctor will ask that the patient does not eat or drink for at least four hours beforehand so that the GI system is as empty as possible. Once in the endoscopy unit, local anesthetic spray is squirted onto the back of the throat to help prevent discomfort as the endoscope is passed down. The patient can choose to be lightly sedated; although this does not induce sleep, it means that most patients retain no memory of the procedure. After the endoscopy, it is advisable not to eat or drink for an hour as the local anesthetic can affect the ability to swallow properly and food may go down the wrong passage.

ENDOSCOPIC RETROGRADE CHOLANGIO-PANCREATOGRAPHY (ERCP)

Sometimes doctors take upper digestive tract endoscopy a step further so that they can examine the ducts draining the gallbladder and pancreas. This technique is also extremely useful in the treatment of patients with gallstones and bile duct strictures who would otherwise need to undergo major surgery.

How does it work?

ERCP combines endoscopic and radiological techniques to provide clear images of biliary and pancreatic ducts and is generally carried out in the X-ray department of a hospital. For this procedure, the patient is given sedation, which induces a light sleep. It is unusual for the patient to remember the procedure afterwards.

Before the ERCP takes place, the patient fasts for about four hours to make sure that the stomach and duodenum are completely empty. The anatomy of the bile ducts can make the procedure technically difficult—to make it easier, the patient lies stomach down on the table, with shoulders turned to one side. Throughout the procedure oxygen is given, and pulse and blood oxygen levels are monitored using a pulse oximeter attached to the finger. The doctor passes an endoscope through the mouth and esophagus into the stomach and duodenum. A long, thin catheter is then threaded down one of the channels in the endoscope and positioned at the point at which the ducts enter the duodenum (see page 28).

Once the catheter is in position, a radioopaque iodine dye that absorbs X-rays is injected back into the ducts—hence the procedure is termed "retrograde"—and a series of X-ray images is taken. Two different types of dye are used: a low contrast one for the bile duct, so that the dye does not obscure gallstones that may be present, and a higher contrast dye for the pancreatic duct. If the doctor finds gallstones or a narrowing in any one of the ducts, it is possible to use the endoscope to carry out treatment there and then (see page 124).

After an ERCP investigation the patient is often able to go home the same day. If a treatment has been carried out, however, an overnight stay in hospital may be necessary.

ON THE CUTTING EDGE

Ingestible camera capsules

It may soon be routine for doctors to examine the insides of the digestive system with a wireless endoscopic camera capsule the size of a vitamin pill. The disposable capsule consists of a battery, a camera on a chip, a light source, and a transmitter. A patient will simply swallow one and continue with normal activities. The camera is "live" for five to six hours and relays information to a video-recording device worn around the waist. Peristaltic movements propel it through the digestive tract, and it leaves the system after 10 to 48 hours.

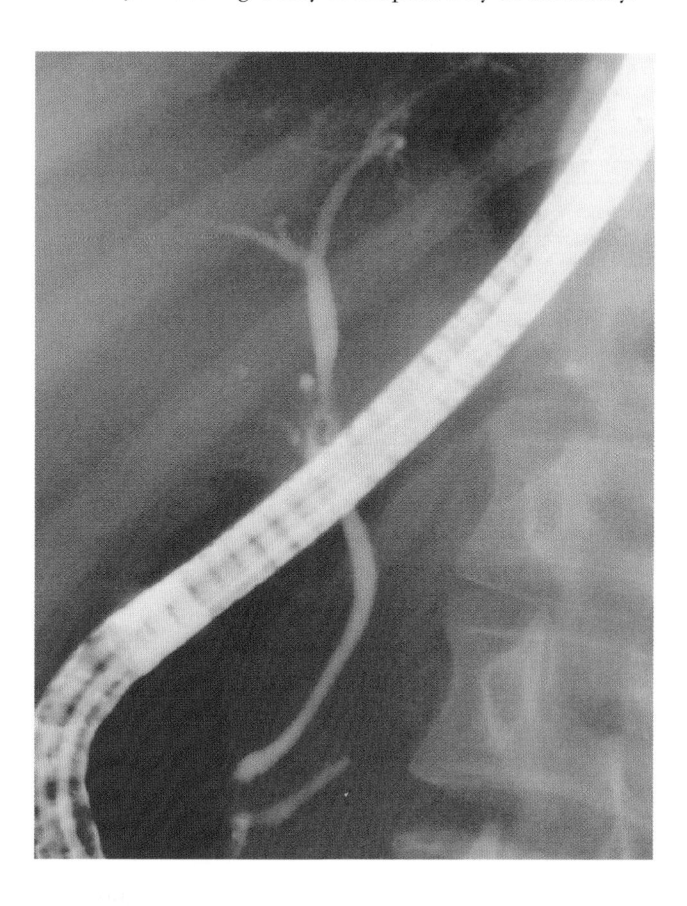

Imaging the parts that other scopes cannot reach

Running diagonally across this contrast X-ray is an endoscope. Visible behind it is the bile duct, filled with radioopaque dye. Running down the right of the picture is the spine.

EGD—looking inside the stomach

In the late 1960s the introduction of techniques using flexible, maneuvrable endoscopes revolutionized the diagnosis of gastrointestinal disorders. EGD (esophagogastroduodenoscopy) is one such technique, providing a quick and easy way of investigating problems with the stomach.

Before the invention of flexible endoscopes, only each "end" of the digestive tract—the esophagus and rectum—could be viewed through rigid tubes. Now, however, doctors can see almost the entire system.

The comfort and cooperation of patients is paramount, so after signing a consent form they are asked if they would prefer to be sedated during the investigation. If so, the doctor administers a sedative, which takes only a minute or so to kick in. The whole procedure is quick: A straightforward gastroscopy takes, on average, about five minutes from start to finish.

The endoscopy team

Two endoscopy staff nurses assist the endoscopist throughout: One talks the patient through the procedure and monitors the pulse rate and blood oxygen levels. The other nurse helps with procedures, taking biopsies and cleaning the scope once the procedure is finished. The unit is equipped with two or three scopes to allow time between patients for disinfection.

Home within 30 minutes

Most endoscopic procedures are carried out on an outpatient basis. Patients who choose to have the procedure without a sedative can go home within 20 or 30 minutes after having something to drink, such as a cup of tea. Those who opt for a sedative have to wait for 45–60 minutes before being taken home by someone else, such as a friend or family member.

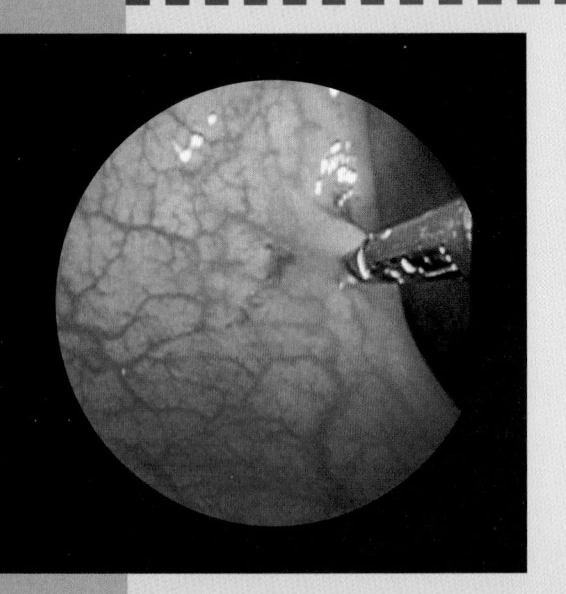

Taking a biopsy

If a growth or other abnormality is found during an examination, the doctor can insert biopsy forceps into a special instrument channel and take a small sample of tissue or cells. The biopsy is brought out through the same channel, and the sample is sent to the pathology laboratory, where technicians can examine it to determine whether the growth is malignant.

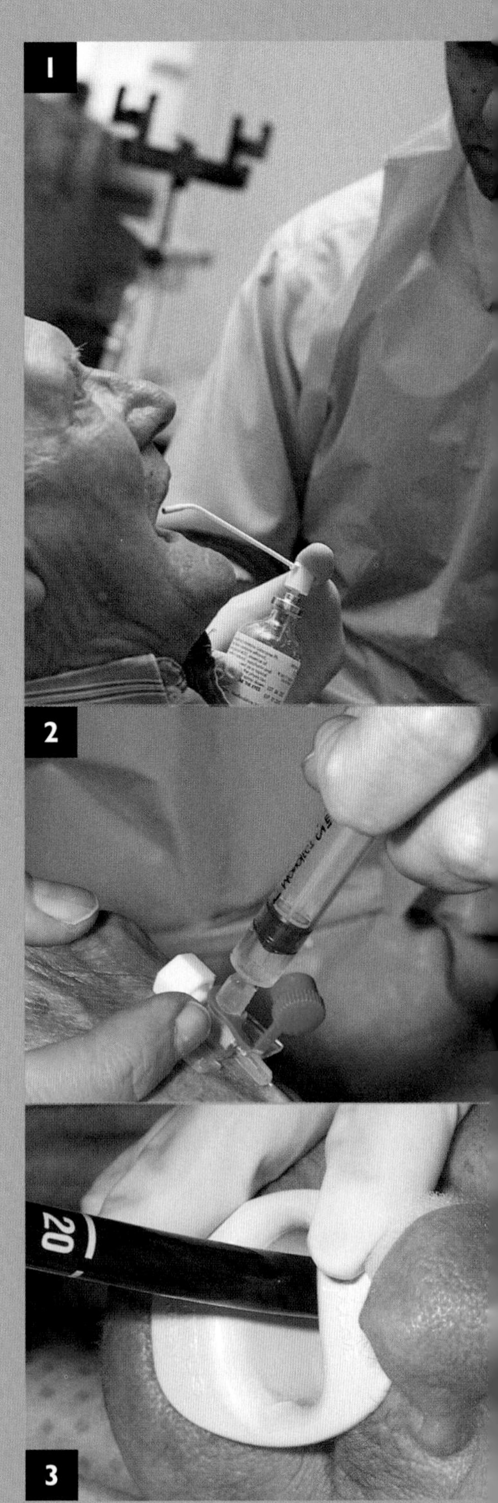

The EGD procedure, step by step

1 The endoscopist sprays some local anesthetic on the back of the patient's throat to numb the area and make the passage of the endoscope easier and less uncomfortable. The patient is then positioned on his left side.

2 The doctor will, if requested, administer a small dose of sedative through a vein in the back of the hand. The nurse attaches a pulse oximeter, which looks like a high-tech "peg," to the patient's forefinger; this device measures how much oxygen is in the blood and takes the pulse throughout the procedure.

3 A mouth guard is put in to prevent biting of the endoscope. To make it easy to breathe while the endoscope is in the mouth, a tube supplying oxygen may be gently inserted into one nostril.

4 The endoscopist first lubricates the endoscope with KY jelly and then gently guides it into the mouth, down the esophagus and into the stomach, at the same time observing the monitor. Wheels on the handset allow the doctor to move the endoscope around to examine the internal surfaces; it can even bend right back on itself to view the top of the stomach.

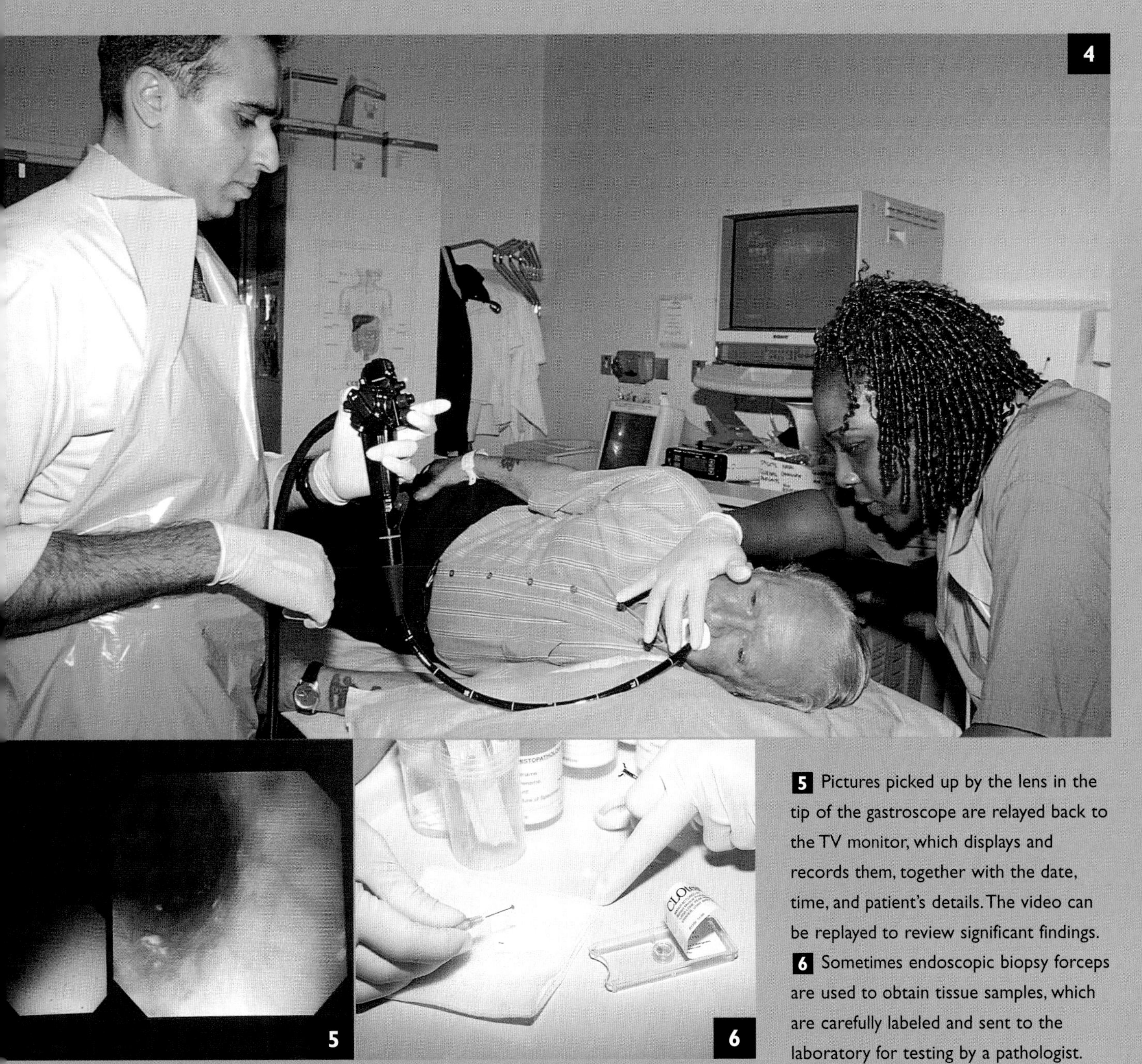

5 Pictures picked up by the lens in the tip of the gastroscope are relayed back to the TV monitor, which displays and records them, together with the date, time, and patient's details. The video can be replayed to review significant findings.

6 Sometimes endoscopic biopsy forceps are used to obtain tissue samples, which are carefully labeled and sent to the laboratory for testing by a pathologist.

EXAMINING THE LOWER DIGESTIVE TRACT

Just like the upper digestive system, the lower GI system can be examined in sections, depending on the symptoms.

- Proctoscopy provides information about the rectum only. It is often carried out if the patient has noticed bright red blood in the stools, which can be an indication of a problem, such as hemorrhoids, in the very last part of the GI system.
- Sigmoidoscopy is used to look at the left side of the colon only and is usually done to investigate persistent diarrhea or someone who has noticed blood in the stools.
- Colonoscopy enables doctors to carry out an examination of the colon from the rectum to the small bowel and is a very effective tool for diagnosing all kinds of digestive system disorders.

For most types of lower digestive tract endoscopy, the doctor carrying out the procedure will ask the patient to lie on his or her left side. This is because the bowel curves to the left, so it is easier and more comfortable to insert the endoscope in this position.

How do they work?

Both rigid and flexible endoscopes are used to look at the lower GI system. The proctoscope is a short, rigid tube about 6 inches long. A rigid sigmoidoscope is a plastic tube with a light source and a magnifying glass at one end, it is used to examine the lower 10 inches of the GI system only. The flexible sigmoidoscope is able to turn corners and can therefore extend to about 28 inches. Colonoscopes are similar to flexible sigmoidoscopes but longer.

Colonoscopes can be up to 5 feet in length.

Proctoscopy

This "on the spot" examination can be done with no preparation at all, either in an outpatient clinic or in the doctor's office. A proctoscope with an obdurator—a central piece that makes insertion more comfortable—is lubricated and inserted through the anus. The obdurator is removed and air is pumped in to inflate the passage for a better view. The doctor then shines a light down the tube to examine the rectum. This investigation takes only a few minutes and is not painful unless there are tender or inflamed areas in the rectum.

Sigmoidoscopy

This technique can be rigid or flexible, depending on the type of endoscope used. Either procedure is slightly more complex than proctoscopy and is usually performed on a hospital ward or in the outpatient clinic. They require little preparation, but sometimes an enema is given to clear the lower end of the GI system. Sigmoidoscopy takes slightly longer than proctoscopy because it views a larger part of the GI system. The endoscope is gently inserted

HAVING A SIGMOIDOSCOPY

The first step in the procedure was an enema, following which the doctor examined my rectum with his finger.

I was a bit apprehensive about this but the doctor was very reassuring. He explained that he had to check that the passages were clear for the scope, and also that it would help to relax the sphincter. Once he'd given the all-clear, the nurse passed him the sigmoidoscope, already coated with KY jelly. I was a little uncomfortable when it was inserted, but the doctor did everything very gently as he guided the tip past the bends in my colon.

At the first bend, there was a tiny stab of pain, but it quickly passed. The worst thing was a feeling that I needed to go to the bathroom for almost the whole duration of the procedure. In just a few minutes the whole tube was in, and the doctor and I watched on the TV monitor as he slowly withdrew it. He assured me that everything looked normal and healthy.

Some blood appeared as the camera got near the rectum, which the doctor said was caused by an anal fissure. At the same time, there was a raspberry sound, as some of the air that had been pumped in leaked out. Then the scope was out, less than ten minutes after we started.

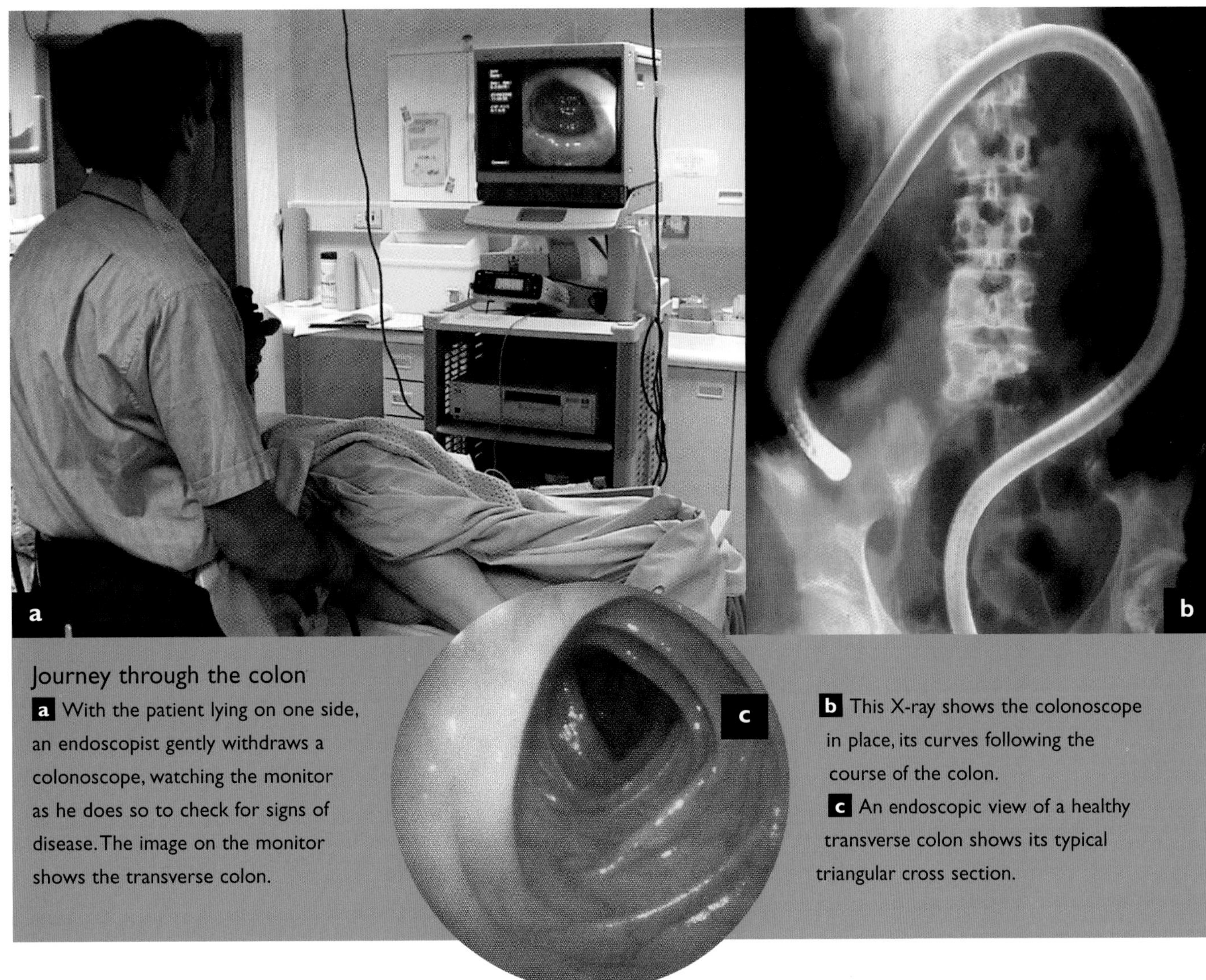

Journey through the colon

a With the patient lying on one side, an endoscopist gently withdraws a colonoscope, watching the monitor as he does so to check for signs of disease. The image on the monitor shows the transverse colon.

b This X-ray shows the colonoscope in place, its curves following the course of the colon.

c An endoscopic view of a healthy transverse colon shows its typical triangular cross section.

into the bowel and then gradually pulled back as the doctor looks through the magnifying glass. Side effects for the patient can include gas and tenesmus—a sensation of incomplete emptying of the bowels.

Colonoscopy

The most important part of a colonoscopy is the preparation to clear the GI system, without which the test would be impossible to carry out. Different hospitals may vary, but patients are usually given the following guidelines.

- Follow a low-fiber diet for two days before the test.
- Drink clear fluids only on the day before the test.
- The day before the colonoscopy, drink the packets of laxative that have been provided with plenty of water.
- Stay close to a toilet after taking the laxatives, as they cause frequent watery bowel movements.
- On the day of the test, don't eat or drink anything for a few hours before the test.

A colonoscopy is usually performed with light sedation, which causes drowsiness—most people are unable to remember anything about the examination afterwards. Oxygen is given throughout, and the pulse and oxygen concentration in the blood are measured continuously using a pulse oximeter.

As the doctor gently advances the endoscope along the intestine, images of the lining of the intestine are relayed to a nearby screen, and any abnormal areas that are spotted are investigated more closely. If the doctor finds something of concern, particularly a growth of some sort, a small sample of tissue—a biopsy—will be taken for laboratory examination. This is done by passing biopsy forceps down a special channel in the endoscope and snipping away a tiny piece of the lining (see page 112).

CURRENT TREATMENTS

Drugs have revolutionized the treatment of some digestive disorders. Peptic ulcers, for example, can now be cured by a simple triple-drug regimen of antibiotics and ulcer-healing drugs. Surgery has also advanced to the stage where a number of procedures can be carried out by keyhole techniques, in which only tiny incisions in the body are made. For some serious conditions, open surgery is still needed, but the risks that go with such operations have been minimized with advances in surgical practices.

Drugs for the digestive system

Drugs are now available to tackle most digestive disorders. Even where an outright cure is not possible, for example with inflammatory bowel disease, drugs can help to alleviate symptoms.

DRUGS FOR NAUSEA AND VOMITING

Vomiting is a reflex action for getting rid of harmful substances, but it may also be a symptom of disease. Nausea and vomiting can be caused by GI infection, travel sickness, or pregnancy or vertigo and can also be a side effect of some drugs, such as those used in chemotherapy. Doctors try to diagnose the cause of sickness before prescribing drugs so that they can focus on treating the underlying cause. Although several drugs for nausea and vomiting—antiemetics—are available without prescription, they should not be taken for more than one to two days without consulting your doctor.

How do they work?

Antiemetics act directly or indirectly on the vomiting center in the brain. The vomiting center is activated by stimuli from several areas of the body, including the chemoreceptor trigger zone in the brain, which is sensitive to chemicals in the blood; the hearing and balance center in the inner ear; and the GI system. The vomiting center and chemoreceptor trigger zone have receptors for several brain chemicals, which act as messengers to signal the sickness response. Antiemetic drugs act on receptors at one or more of these sites.

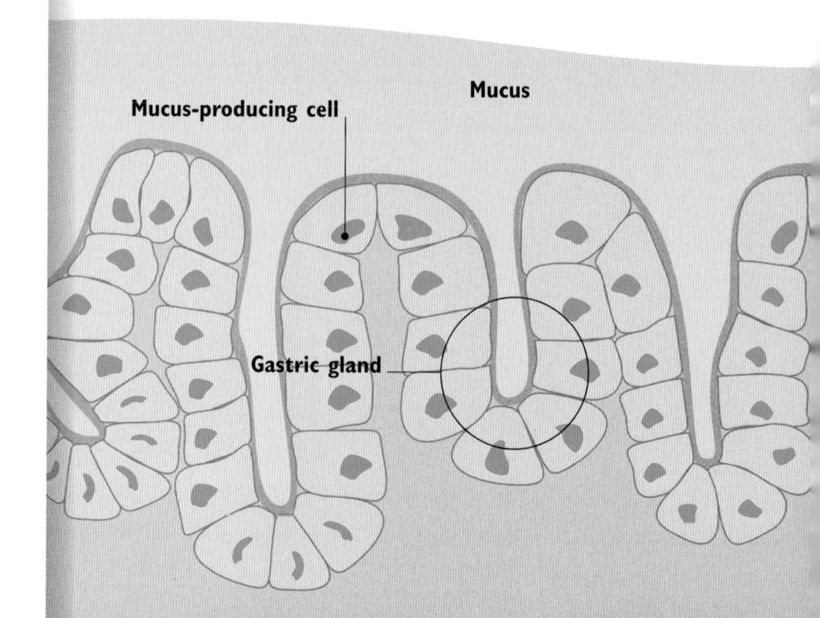

- Antihistamines These drugs, such as meclizine, work at sites both in the inner ear and on the vomiting center. They can be bought over the counter and are useful in preventing and treating motion sickness and dizziness.
- Hyoscine hydrobromide This is used to control motion sickness. It may be applied in the form of a tiny patch behind the ear to allow continuous release of the active ingredient.
- Dopamine receptor antagonists Domperidone and metoclopramide, two examples of this group, block receptors in the chemoreceptor trigger zone. They are often prescribed to prevent and treat chemotherapy-induced nausea and vomiting.
- Phenothiazines Drugs in this group include prochlorperazine, which blocks receptors in the chemoreceptor trigger zone. They are used to prevent and treat nausea and vomiting in people with cancer and in those taking drugs that cause nausea.
- Serotonin receptor antagonists These block receptors for the brain chemical serotonin and are effective against nausea and vomiting induced by anticancer drugs. An example is ondansetron.

Stomach acid is incredibly strong—if it came into direct contact with your skin, it would burn it.

What are the adverse effects?

Some antiemetics, including prochlorperazine, meclizine, and hyoscine hydrobromide, may cause drowsiness and shouldn't be taken if driving or operating machinery. Some can also cause a dry mouth, blurred vision, or difficulty in passing urine, or all three.

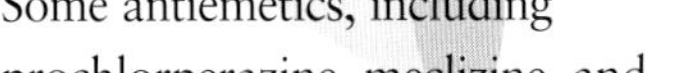

ANTACIDS

Antacids are one of the most widely used classes of drugs. People take antacids mainly to relieve indigestion or heartburn, known medically as dyspepsia. Often a change in diet or a glass of milk can prevent or relieve indigestion, but many people find antacids a huge help from time to time. Doctors often prescribe antacids for mild symptoms of dyspepsia in disorders such as inflammation or ulceration of the esophagus, stomach, and duodenum. They are usually taken when symptoms occur and relieve pain within a few minutes. The drugs may be prescribed by a doctor, but more often people buy them from a pharmacy or supermarket. Antacids are available as tablets (some of which are chewable) and liquids.

How do they work?

The stomach lining is normally protected from the powerful stomach acid by a thin layer of mucus. There are several types of antacid—all are simple chemical compounds that are mildly alkaline and so help to neutralize stomach acid. Taking antacids on a regular basis reduces the acidity of the stomach, which is good for healing ulcers, but can be bad for digestion of food.

- Aluminium and magnesium compounds Both these groups of antacids have a prolonged action in neutralizing stomach acid and are widely used for the treatment of indigestion.

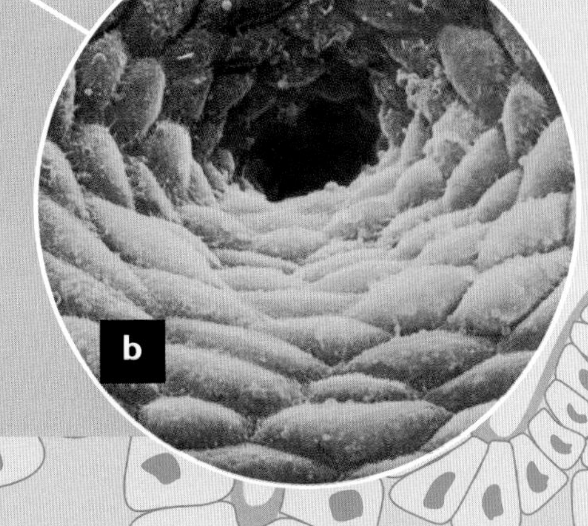

Shield of mucus

The stomach produces two to three quarts of highly acidic gastric juice per day. In order to prevent the stomach from being eaten away by its own juices, cells at its surface produce a thick coating of mucus that forms a protective barrier.

a An electron micrograph of the stomach lining shows the entrances to gastric glands, which secrete hydrochloric acid. The actual surface of the lining is made of simple columnar cells, which produce mucus.

b A close-up shows the mucus-producing cells spiralling down into a gastric gland.

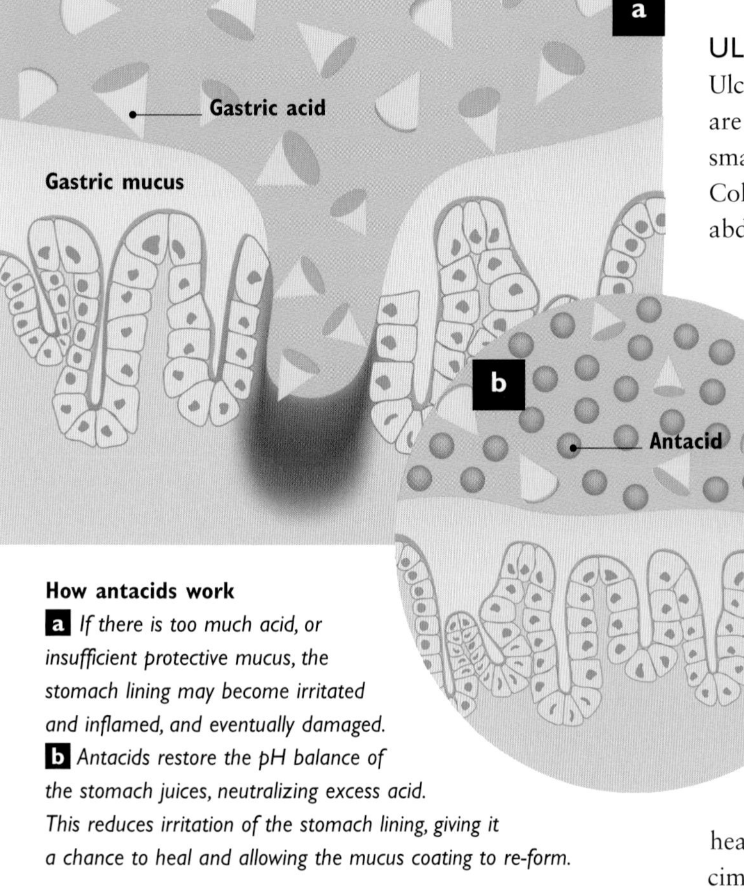

How antacids work

a *If there is too much acid, or insufficient protective mucus, the stomach lining may become irritated and inflamed, and eventually damaged.*

b *Antacids restore the pH balance of the stomach juices, neutralizing excess acid. This reduces irritation of the stomach lining, giving it a chance to heal and allowing the mucus coating to re-form.*

- Sodium bicarbonate This antacid works rapidly to neutralize stomach acid, but its action is short-lived.
- Combined preparations Some antacids are combined preparations: a neutralizing compound, such as magnesium, along with other substances called alginates. Alginates float on the contents of the stomach and produce a neutralizing layer that prevents acid from rising into the esophagus and causing heartburn.

What are the adverse effects?

Aluminium compounds may cause constipation, and magnesium can cause diarrhea; the two may be taken together to counteract their opposing effects. Aluminium compounds can also cause weakness and bone damage, and should not be taken at high doses for long periods.

Individuals with kidney problems should take extra care because levels of magnesium in the blood may become high, causing lethargy and drowsiness. Sodium bicarbonate produces gas and so may cause bloating and belching. People with heart or kidney diseases should not take sodium bicarbonate because it can lead to water accumulation and chemical imbalances of the blood.

ULCER-HEALING DRUGS

Ulcers most commonly form in the stomach, where they are called gastric ulcers, and in the duodenum (part of the small intestine), where they are called duodenal ulcers. Collectively known as peptic ulcers, they can cause abdominal pain, vomiting, and changes in appetite. Left untreated, ulcers may erode blood vessel walls or perforate the stomach or duodenum, with potentially fatal consequences. Although the symptoms of an ulcer may be relieved by antacids, ulcer healing is slow. Instead, the usual treatment is with a specific ulcer-healing drug. Some of these drugs are combined with antibiotics to treat *Helicobacter pylori* infection—now known to be the main cause of peptic ulcers (see page 122).

How do they work?

Most ulcer-healing drugs dramatically reduce acid secretion in the stomach. The result is that they both relieve symptoms and heal the ulcers. Some ulcer-healing drugs, such as cimetidine, can be bought over the counter. But since they may mask the symptoms of a more serious illness, such as stomach cancer, they should not be used for more than two weeks without consulting your doctor.

Milestones IN MEDICINE

Until 1982 the cause of peptic ulcers was uncertain, but overproduction of acid was thought to be responsible. Then researchers discovered a bacterium—*Helicobacter pylori*—that could survive in acidic stomach conditions by burrowing into the mucus. There was still some doubt as to whether this bacterium caused ulcers, so to clarify the issue, an Australian doctor, Barry Marshall, decided to infect himself with it. He developed gastritis, and a biopsy confirmed *Helicobacter pylori* as the culprit. This brave experiment led to antibiotic treatment for ulcers.

- **Histamine (H_2) receptor blockers** These drugs have structures that fit the histamine (H_2) receptors on the acid-secreting cells in the stomach. Histamine normally attaches to these receptors and stimulates acid secretion. It also mediates the action of other substances, such as gastrin and acetylcholine, which further increase acid production. H_2 receptor blockers prevent histamine from attaching to its receptor site and so reduce acid production, allowing the mucus lining to heal. Examples are cimetidine, ranitidine, famotidine, and nizatidine. They are usually taken for four to six weeks.
- **Proton pump inhibitors** Acid secretion by cells in the stomach depends on the action of a group of enzymes that work together in a system known as a proton pump. Drugs that bind to the proton pump enzymes are known as proton pump inhibitors and include omeprazole, lansoprazole, and pantoprazole. They can reduce acid secretion by 90–100 percent over 24 hours. Proton pump inhibitors are effective for esophagitis (inflammation of the esophagus) and esophageal and peptic ulcers. They are particularly useful for ulcers that have failed to heal after other treatments. They begin to reduce pain in a few hours and usually allow the ulcer to heal within three to eight weeks.
- **Misoprostol** This drug is related to naturally occurring chemicals called prostaglandins. It reduces the amount of acid secreted in the stomach and promotes healing of peptic ulcers. Drugs such as NSAIDs have an antiprostaglandin action, which blocks certain prostaglandins and causes ulcers and bleeding in the digestive tract. Misoprostol can be tried to prevent or cure such ulcers—treatment is sometimes effective within a few weeks.
- **Sucralfate and bismuth** Sucralfate forms a coating over the ulcer, protecting it from the action of the stomach acid and allowing it to heal. Bismuth may stimulate production of GI-protective prostaglandins or bicarbonate, which contribute to the protective mucus layer of the stomach. It also kills *H. pylori* bacteria and is used in triple therapy with antibiotics.

A closer look at stomach ulcers

a This colored electron micrograph of the surface of the stomach lining shows a smooth, round wound—an ulcer—in the otherwise pitted epithelial surface.

b This micrograph shows a section through a sample of tissue taken from a bleeding ulcer in the stomach. The mucus-secreting cells are stained green and seem fairly normal on the right, but toward the left-hand side of the image, the cells are contorted with massive numbers of red blood cells.

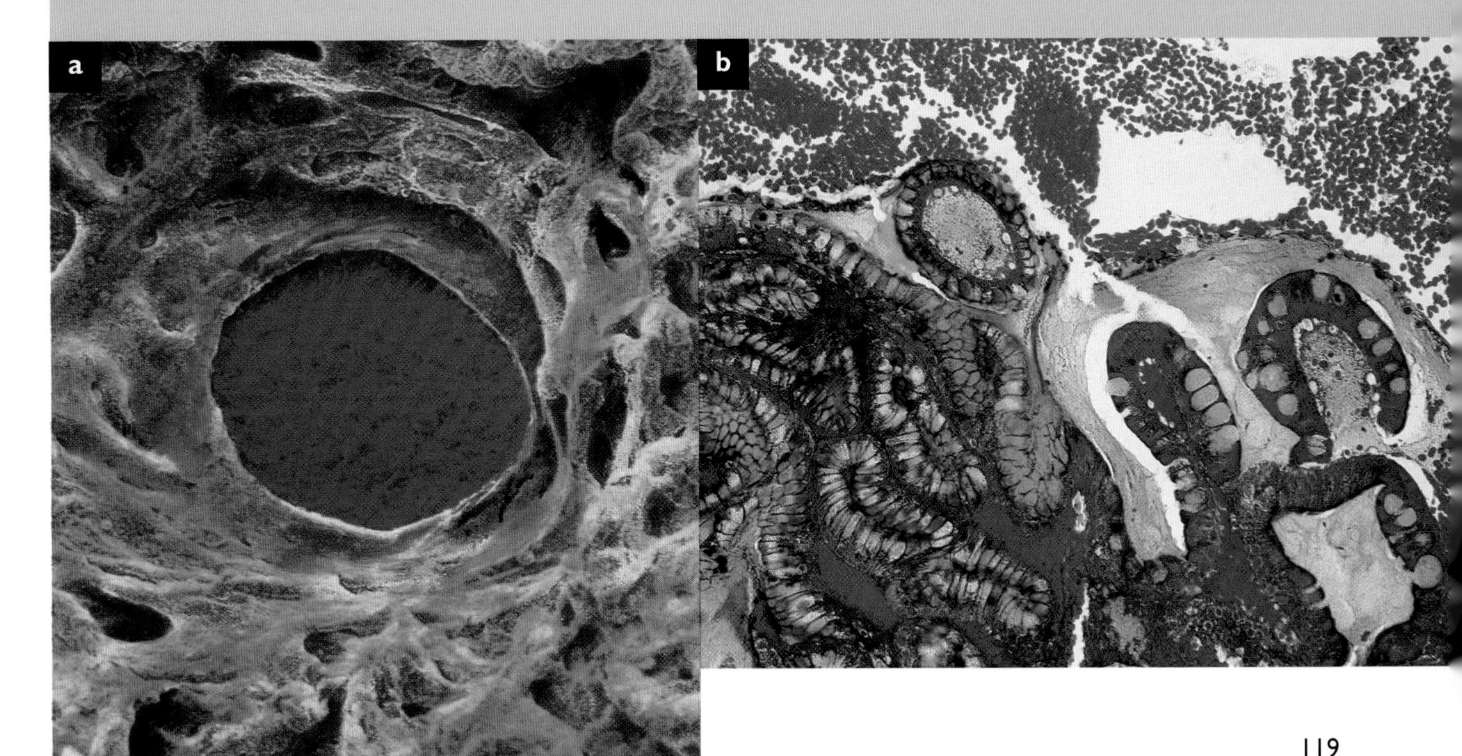

What are the adverse effects?

H_2 receptor blockers can cause confusion in elderly people. On rare occasions, cimetidine can cause temporary breast enlargement and impotence in men. Proton pump inhibitors may cause headaches, diarrhea, and rashes and shouldn't be taken by anyone with liver problems or by pregnant or breastfeeding women. Many people find they experience side effects (nausea, vomiting, and a rash) with triple therapy using proton pump inhibitors and antibiotics, mainly related to the antibiotics. The most likely side effects from misoprostol are diarrhea and indigestion. If they are severe, it may be necessary to stop treatment and find an alternative. Sucralfate can cause diarrhea or constipation as well as nausea and indigestion.

DRUGS FOR DIARRHEA

An attack of diarrhea usually resolves quickly without medical attention. The most important principle is to abstain from food and drink plenty of clear fluids. Rehydration solutions (containing sugars as well as potassium and sodium salts) are available from pharmacies and can be used to combat dehydration and chemical imbalances caused by lost body fluids. Alternatively, you can make your own (see below).

Rehydration for diarrhea
This easy-to-make recipe delivers optimum oral rehydration. In 2 pints of drinking or boiled water, add 2 tablespoons of sugar or honey, ¼ teaspoon of salt, and ¼ teaspoon baking soda. For a dose of potassium, add some mashed banana.

Diarrhea bug
These Clostridium *bacteria belong to one of the many species of microorganism that can cause diarrhea.*

Several drugs are available to relieve non-specific diarrhea when the diarrhea is particularly severe and once it is certain that the diarrhea is neither infectious nor toxic. These drugs include opioids and bulk-forming and adsorbent agents. Antispasmodic drugs may also be used to relieve cramping pain. If diarrhea is found to be caused by an infectious agent, and the agent can be identified, a course of antibiotics may be prescribed (see page 122).

How do they work?

Antidiarrheal drugs act on the gut wall, affecting the speed with which it moves contents along or by boosting the large intestine's ability to absorb water, thereby making the stools less fluid.

- Opioid drugs The commonly used drugs are loperamide, codeine, and co-phenotrope. These drugs act directly on the GI wall and slow down GI motility. Feces then pass more slowly through the large intestine. They are very effective and are used when the diarrhea is severe and debilitating. Some of these drugs—for example, Imodium (which contains loperamide)—can be bought over the counter.
- Bulk-forming agents These agents have a milder effect than opioid drugs and are usually used only when it is necessary to regulate bowel action over a prolonged period, especially in individuals who have had part of the intestines removed by ileostomy or colostomy (see pages 130–131). The agents contain particles that swell up as they absorb excess water, irritants, and harmful chemicals from the intestines, making the feces firmer and less fluid.
- Antispasmodic drugs These are anticholinergic drugs—drugs that act on the involuntary part of the nervous system—and include hyoscine. They are used mainly for the symptoms associated with irritable bowel syndrome. They reduce the transmission of nerve signals to the large intestine wall, thus preventing spasms.

What are the adverse effects?

All antidiarrheal drugs should be taken with plenty of water and may cause constipation if used in excess. It is important not to take a bulk-forming agent together with an opioid or antispasmodic drug because a bulky mass could form and obstruct the large intestine.

LAXATIVES

The simplest remedy for constipation is to drink more fluid, eat a high-fiber diet, and get more exercise. Sometimes, however, laxatives may be required. Laxatives are also used to prevent pain and straining in people suffering from hernias or hemorrhoids. Doctors may prescribe laxatives for the same reason after childbirth or abdominal surgery. Potent laxatives are used to clear the large intestine before investigative procedures such as a barium enema (see pages 106–108) or colonoscopy (see page 115). Although laxatives can be bought over the counter, anyone taking them regularly should consult a doctor, because they may have adverse effects if taken inappropriately.

Average daily stool weight in the West is 1¾–7 ounces, whereas in Africa it is about 17½ ounces.

How do they work?

Most laxative drugs work in one of three ways: they stimulate the intestinal wall to move the contents along faster, they help draw fluid into the feces, or they bulk up the feces to make stools easier to pass.

- **Bulk-forming agents** Ispaghula husk and methylcellulose take a while to work but are less likely than other laxatives to interfere with bowel action. These agents are usually taken after meals but are not absorbed as they pass through the intestines. They contain particles that absorb many times their own volume of water to bulk-up the feces and so encourage normal bowel action. They should be taken with plenty of fluids.
- **Stimulant (contact) laxatives** These substances, including senna and bisacodyl, act on the nerve endings in the intestinal wall that trigger muscle contraction. Feces are moved through the large intestine faster so there is less time for water to be absorbed by the body. The feces become more liquid and are passed more easily.
- **Osmotic laxatives** These laxatives act by increasing the amount of water in the large intestine. Feces absorb the water with the result that stools become softer and more bulky. Osmotic laxatives are available as tablets or as suppositories, which tend to act more quickly. This group includes Epsom salts (magnesium sulphate), given before surgery to clear the bowels.

What are the adverse effects?

In general, laxatives should be used only for short periods of time. Prolonged use may be harmful and may render an individual dependent on them for regular bowel motions. The colon can become "lazy" and unable to contract normally. Long-term use of stimulant laxatives can cause abdominal cramps and diarrhea; osmotic laxatives used long term can cause chemical imbalances in the blood.

DRUGS FOR INFLAMMATORY BOWEL DISEASE

Inflammatory bowel disease is an abnormal immune reaction and includes Crohn's disease and ulcerative colitis. Two main groups of drugs are used to treat these conditions.

How do they work?

Both drug groups come in tablet, enema, and suppository forms, and they work by damping down the body's natural inflammatory response, although they do this in different ways.

TALKING POINT

Do-it-yourself prescribing

Mild bowel discomforts are so common as to be almost normal—the average adult suffers about one symptom a day. It is neither possible nor desirable to go to the doctor for every minor ache and pain, yet it is also entirely reasonable not to want to suffer unnecessarily. This means that self-care with home remedies and medicines that are available over the counter is an important part of health care. But can you overdo it? Almost certainly the answer is yes, and it's probably wise to talk over your symptoms with a friend or relative before you take even as much as a simple antacid. If you find you need to self-medicate on a regular basis, you should visit your doctor.

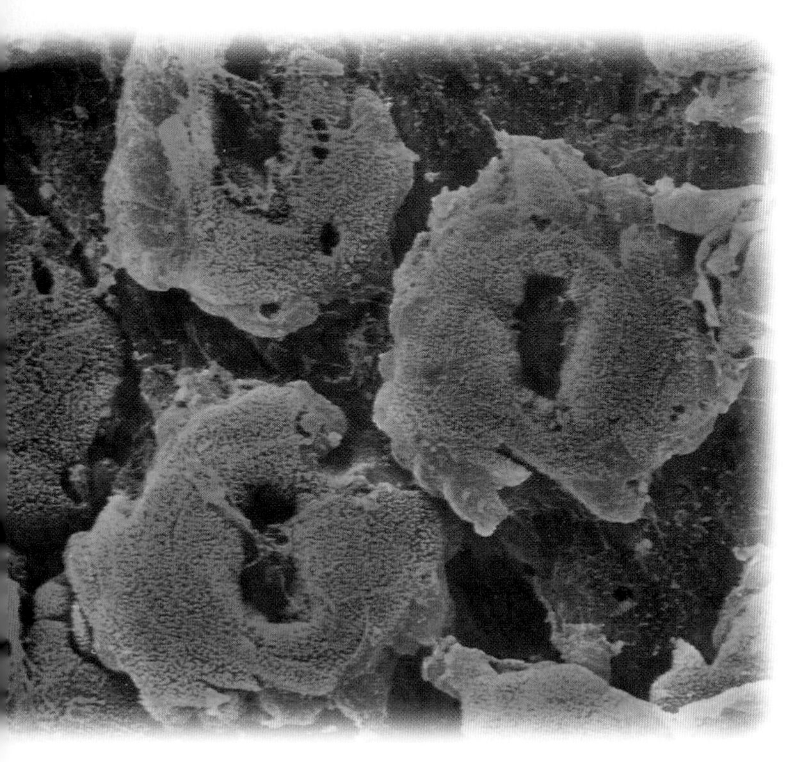

Inflammatory bowel disease
This false-colored micrograph shows the colon of an ulcerative colitis sufferer, magnified 1100 times. The surface of the colon (pink) has been eroded, leaving the mucous glands (orange) abnormally raised.

- Aminosalicylates These drugs, which include sulphasalazine and mesalazine, prevent chemicals called prostaglandins from being released by the damaged lining of the intestine. Prostaglandins are part of the body's inflammatory response. Aminosalicylates are used for acute attacks of colitis (inflammation of the colon), but their main role is to prevent recurrent attacks.
- Corticosteroids These drugs act to prevent movement of white blood cells into the damaged area. In Crohn's disease or ulcerative colitis, corticosteroids (steroids for short) may be given as suppositories or foam enemas to deliver the drug quickly to the site of action. Steroids can be used as a short high-dose course in severe attacks of inflammatory bowel disease.

What are the adverse effects?

Aminosalicylates can cause diarrhea, nausea, and headaches. There is also a small risk of blood disorders and people taking this drug are always told to report any unexplained bruising, bleeding, or sore throats. Adverse effects of long-term use of corticosteroids include diabetes, osteoporosis (thinning of the bones), and indigestion. In children, corticosteroids can cause growth retardation. In adults, continuous high-dose usage may cause Cushing's syndrome—characterized by a moon face, acne, bruising, edema (swelling), and fat deposition on the back. Anyone on long-term steroid therapy should carry a card giving details of the drug and dosage.

While a person is on long-term oral steroid therapy, the natural production of corticosteroids by the adrenal glands is suppressed. Withdrawal of steroid treatment should be gradual and in consultation with a doctor, to allow the glands to resume production of natural steroids.

ANTIBIOTICS

The use of antibiotics in gastrointestinal disease is confined to specific infections, and they are only given when the infective cause is known.

Infection with the bacterium *Helicobacter pylori* is present in nearly all patients with duodenal ulcers and in about 75 percent of patients with gastric ulcers. For this reason, antibiotics are now used as part of the treatment for ulcers. Combining an ulcer-healing drug with two antibiotics has a major impact on preventing ulcers from recurring—a significant problem when ulcer-healing drugs are used alone. This triple drug therapy is highly effective and needs to be taken for only one or two weeks.

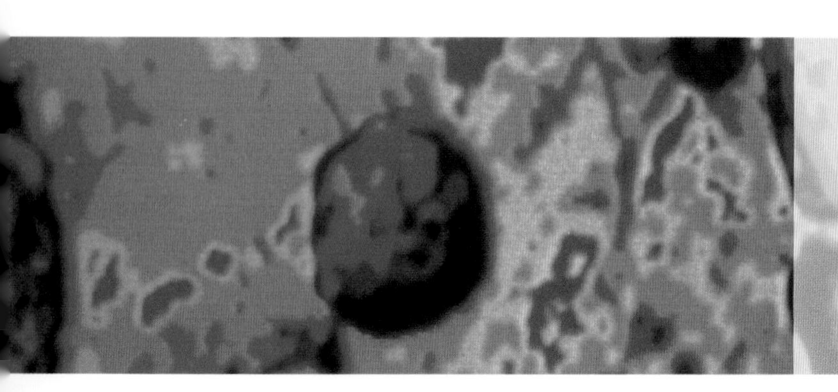

About 50 percent of American adults are infected with* Helicobacter pylori*, significantly increasing their risk of developing ulcers.

How do they work?

Antibiotics act by interfering with metabolic pathways in the bacteria, especially the ones that bacteria use to make their cell walls, so that they disintegrate.

What are the adverse effects?

Antibiotics can cause diarrhea, although it is not usually severe. They can also alter the balance of healthy GI flora and the yeast *Candida albicans*. Thrush, a vaginal infection caused by this yeast, is a fairly common side effect of antibiotic treatment in women.

DRUGS FOR ANAL AND RECTAL PROBLEMS

Several different types of drugs are used in the form of creams, suppositories, or enemas to treat problems such as hemorrhoids (piles) and anal fissures.

How do they work?

Drugs for anal and rectal problems are a diverse group and mainly act as antiinflammatories. These include anesthetic creams and gels, and corticosteroid creams and suppositories.

- **Local anesthetics** These act on nerves supplying the skin and prevent transmission of signals to the brain. Formulations are available as a spray, cream, or gel.
- **Topical corticosteroids** Using an applicator, these can be applied directly to the hemorrhoids, or they can be used in the form of a suppository. They help to reduce inflammation and can relieve itching and discomfort.

What are the adverse effects?

There are no major side effects associated with these agents, but they should not be used for more than a few days without consulting your doctor.

ANTIPARASITIC DRUGS

The human body provides a suitable environment for the growth of many types of parasite, including insects, worms, and flukes. Some can be transmitted from person to person by direct contact, others are picked up through inhalation of infected air or consumption of contaminated food or water. Treatment with antiparasitic drugs is necessary in combination with hygienic measures to break the cycle of infection.

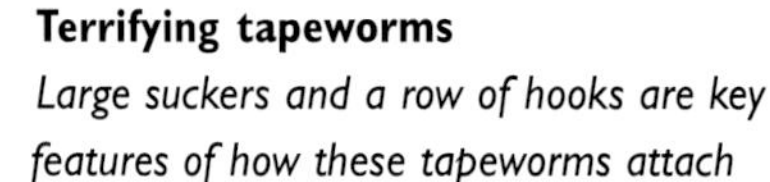

Terrifying tapeworms
Large suckers and a row of hooks are key features of how these tapeworms attach themselves to the intestine. Tapeworms affect some 4 million people worldwide.

Parasitic infections are most common in tropical countries and areas of poor sanitation, but because of more worldwide travel, immigration, and imported foods, parasites are also found in the United States.

How do they work?

Antiparasitic drugs work directly on the parasite by paralyzing it or killing it outright, so that it passes out of the body. Others exert their effect on tissues bodywide.

- **Anthelmintics** Drugs such as mebendazole are used to treat roundworms, tapeworms, and threadworms. They kill or paralyze the worm, which is then excreted.
- **Praziquantel** This drug is used for infection with flukes such as the liver fluke. These parasites live in the bile duct in the liver and can cause jaundice. Praziquantel is active against all species of liver flukes.

What are the adverse effects?

Taken as a single dose or short course, these drugs do not usually produce side effects. However, nausea, vomiting, and abdominal pain can result.

Endoscopic treatments

As well as being used for investigations of the digestive tract, endoscopy can be used to treat particular conditions. Often the endoscopist will carry out any treatment that is needed as an extension of the investigative endoscopic procedure.

One of the most common investigative techniques used in gastroenterology today is endoscopy, using a telescopic device with a built-in camera to see inside the body. Thanks to the unique design of the endoscope, doctors can not only look at the structures of the stomach and small and large intestine, they can also treat many problems. As well as having their own lighting and video-recording systems, endoscopes contain channels down which instruments can be passed for an appropriate task, removing a polyp from the inside of the colon, for example, using polypectomy snares. A suction and water channel helps to clear debris and wash the lens.

PREPARING FOR THE PROCEDURE

At the endoscopy unit, a nurse asks a series of questions to check that the patient is generally healthy and ready for the procedure. To clear the GI system of food, the patient usually has to fast for at least four hours beforehand, take laxatives to clear the bowels, or both. A mild sedative may be administered to relieve anxiety.

Once in the endoscopy suite, the patient lies down on the bed and may be positioned in various poses—usually on the left side—by the nurses or doctors to assist the viewing process. A local anesthetic is sprayed onto the back of the throat, for an upper GI endoscopy, to make the patient more comfortable as the endoscope is inserted into the mouth and down the throat.

The endoscope is fed slowly into the digestive tract so that the doctor can fully inspect the section of the GI system in question using the endoscopic video camera. Once a full examination is completed, the doctor may want to sort out any potential or obvious problems visible during the endoscopy. Gastrointestinal conditions that are routinely treated in this way include

- narrowing of the esophagus;
- bleeding in the digestive tract and from ulcers; and
- tumors or polyps.

Treating digestive system problems with an endoscope avoids open surgery and its associated risks. A general anesthetic is not required, recovery is much quicker, patients experience much less pain, and the chances of infection are generally lower. Endoscopic treatment also means that patients can be treated on an outpatient basis. The procedures are so quick—a gastroscopy, for example, can take as little as five minutes—that a patient can often return home the same day. Endoscopy is easily repeatable, which is important for instance, for monitoring an ulcer.

CORRECTING A NARROWED ESOPHAGUS

Parts of the esophagus can be constricted by conditions such as tumors. This can be corrected using balloons or stents (mesh tubes), passed down the esophagus via the endoscope. A balloon is inflated once in place, widening the narrowed portion, or stricture. In some cases a stent may be inserted to keep the passage from closing again.

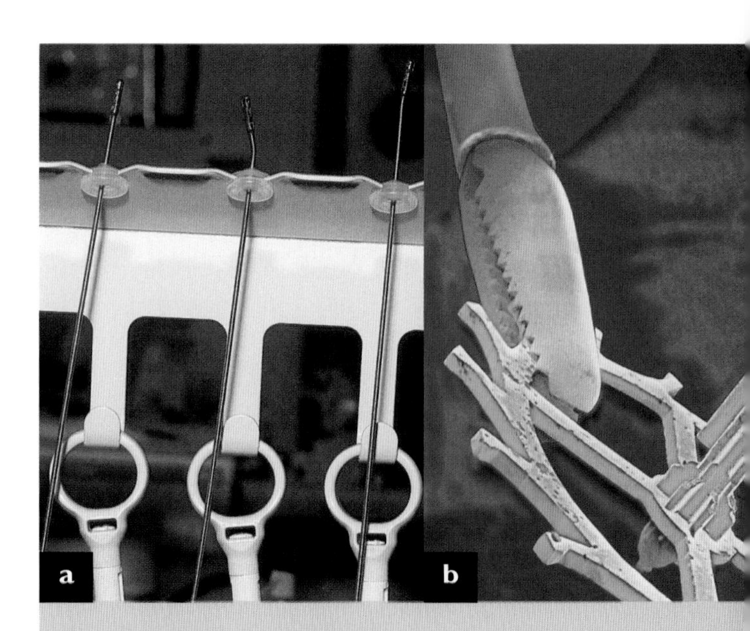

The right tool for the job

a The endoscopist can choose from a wide range of tools, which are fed down a special channel in the endoscope.
b Endoscopic microforceps, magnified 16 times, grasp a watch cog, demonstrating their precision of movement.
c A nurse assists an endoscopist during an EGD procedure. They watch on a video monitor as he guides the instrument down the patient's throat and into the stomach.

STANCHING A BLEEDING ULCER

Endoscopy is routinely used to identify bleeding ulcers, but it can also be used to treat them. Once the endoscopist has located an ulcer, the leaking blood vessels are either cauterized with a heating probe or injected with a diluted epinephrine solution. Epinephrine makes the blood vessels constrict and so slows and then stops the bleeding.

REMOVING A TUMOR OR POLYP

Benign or malignant tumors can be cut out of the digestive tract during endoscopic procedures. Colon polyps, for instance, are generally excised during colonoscopy, either by burning or by cutting them out with a wire loop, known as a snare. It may take more than one procedure to do this if there are numerous polyps or if one polyp is particularly large.

If a condition warrants more extensive treatment, a surgeon may perform open surgery or, for selected problems, surgery using a type of endoscope known as a laparoscope (so-called keyhole surgery—see page 128).

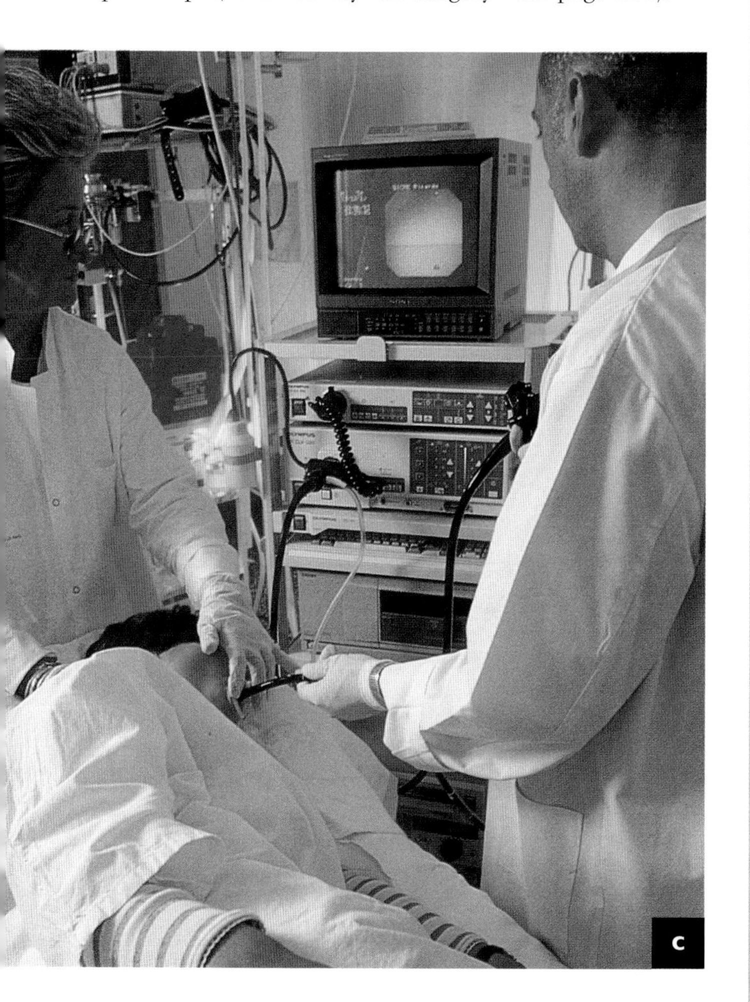

The inside track

Once the matter of science-fiction movies, surgical procedures done from within the body are now routine in the treatment of many digestive tract disorders.

a A narrow tube leading away from the endoscopic camera carries air to a balloon lodged in a narrowed stretch, known as a stricture, farther down the esophagus. Inflating the balloon widens the stricture.

b An endoscopic view of an esophageal stent holding open a stricture. Stents are tubes made out of wire mesh. They are inserted in collapsed form, and their mesh structure allows them to be expanded, like a concertina, once in place.

c Esophageal varices are a serious complication of liver damage, in which high blood pressure causes weak blood vessels in the esophagus to swell and protrude. They can be treated by banding, where rubber bands are placed around the base of the varix using endoscopic manipulation. This endoscopic view shows a large varix in the center of the picture, with a blue band around a varix behind it. Banding cuts off the circulation to the protruding vessel, preventing it from bursting and hemorrhaging.

d Large polyps may be removed during colonoscopy. Here a polyp (right) in the sigmoid colon has been cut away from its base, leaving a small wound (left).

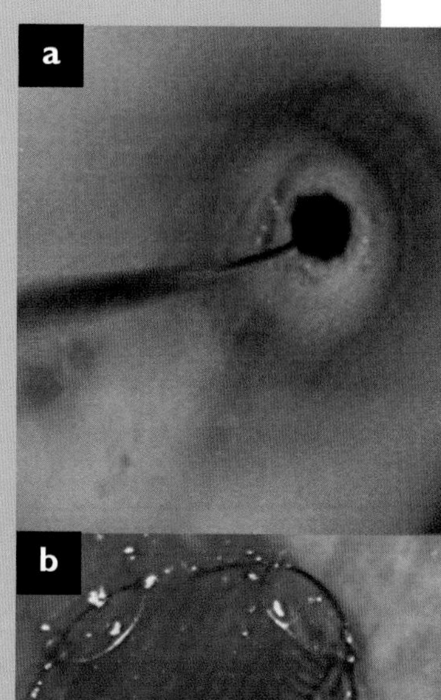
a

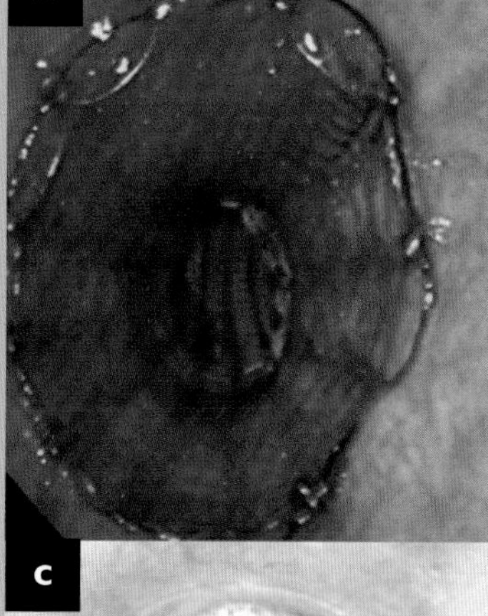
b

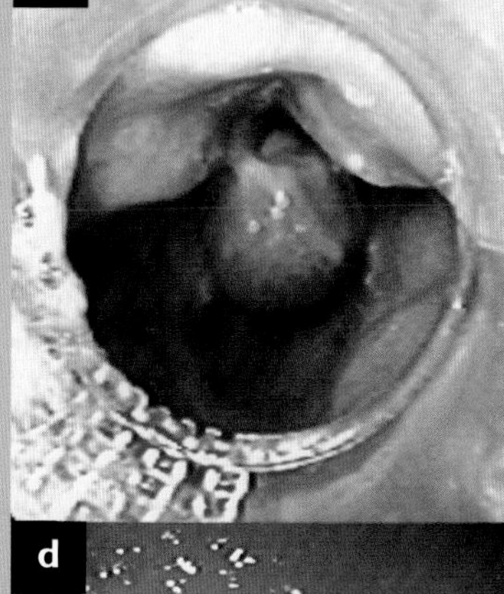
c

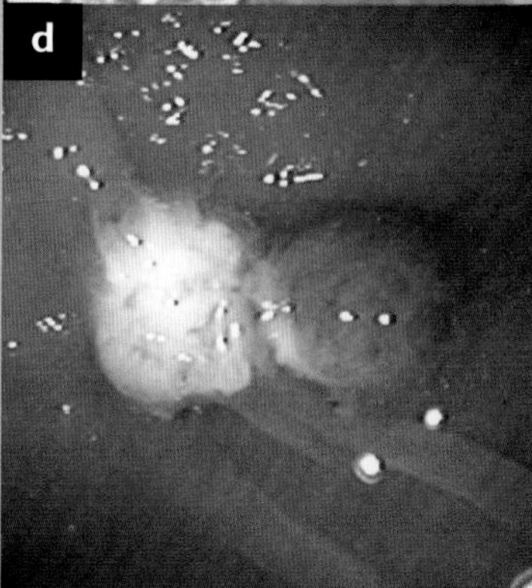
d

Surgical solutions

Surgery plays a major role in gastrointestinal treatment, either as the first-line treatment or after drug therapy has failed. Surgical advances allow some operations to be carried out as a day case. Newer "keyhole" techniques mean fewer complications and faster recovery.

Abdominal surgery ranges from the simple—operations carried out on an outpatient basis—to major procedures in which large amounts of the digestive tract are removed because of disease. Most procedures are planned in advance (elective surgery), but sometimes emergency abdominal surgery is needed.

BEFORE AN OPERATION

For any operation, the patient is examined before surgery and will meet the surgeon performing the procedure. This is the time when the patient can ask questions and give informed consent. For relatively minor operations such as hernia repair, the patient is usually admitted on the day of the procedure and can often go home that same day. If specific tests are needed—for example, full blood count, measurement of urea and electrolytes, chest X-ray, or electrocardiogram (ECG), depending on the patient—patients may have to attend a preadmission clinic a few days before the operation. If the operation requires general anesthetic, it may be necessary to go to the hospital the day before the operation. For major abdominal surgery, especially if this involves opening the intestines, doctors give an injection of antibiotics as part of the premedication an hour or two before the procedure. These antibiotics help to prevent possible infection from the intestine entering the abdominal cavity.

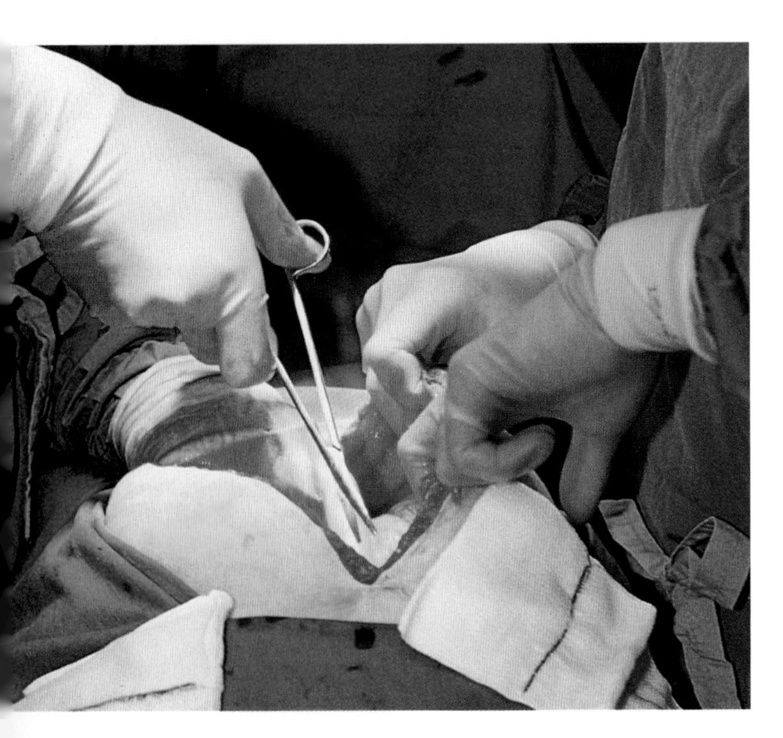

RECOVERING FROM ABDOMINAL SURGERY

After any major surgery, the patient wakes up in the recovery room, often with a number of tubes attached, which provide fluids and pain-relieving medicines. After major abdominal surgery, it can take more than two to three days for complete gastrointestinal function to return. During this time it is not possible to pass gas or stools. After the operation, doctors carry out a daily abdominal examination to check for tenderness and to make sure that the intestines have not become distended due to an obstruction resulting from the surgery. Any surgery on the esophagus makes it impossible to eat for several days, so the patient is fed through a nasogastric tube passed through the nose and into the stomach.

APPENDIX REMOVAL

Appendectomy—removal of the appendix—is the most common abdominal emergency procedure carried out between childhood and early adulthood. The appendix serves no known function in humans, but if this tiny fingerlike organ becomes inflamed, it can burst and cause life-threatening infection. The aim of appendectomy is to remove the inflamed appendix before it bursts.

The operation is carried out under general anesthesia through a small incision (1–2 inches) in the lower right abdomen just above the groin. The appendix is tied at its base with thread then cut off. The whole procedure takes about 30–60 minutes, and the patient can go home after two or three days. Over the next couple of weeks, a patient can gradually resume normal activities.

Open abdominal surgery
Many surgical procedures on the digestive system involve opening up the abdominal cavity, which can require a long incision. The location and length of the cut depends on the organ being operated on.

Keyhole surgery—the way forward?

TALKING POINT

An increasing number of operations are now performed by keyhole (laparoscopic) surgery rather than standard open surgery involving a large incision to reveal the operating site. There are several advantages of laparoscopic surgery, but it is not suitable in every case and for every procedure. The main advantage is that only two or three very small incisions are needed—usually no more than ½–¾ inches long. This not only reduces the risk of infection but also speeds recovery. Plus, there is less scarring. But laparoscopic surgery is often a longer procedure than open surgery. Conventional open surgery is more appropriate in emergency situations where the surgeon has to have rapid and clear access to the operating site.

HERNIA REPAIR

One of the most commonly performed operations on the digestive system is the repair of a hernia. There are several types of hernia, and treatment varies depending on the type and position. The most common is in the groin (an inguinal or femoral hernia), which is caused by a weakness or defect in the abdominal wall that allows the intestines to protrude through. If a small bend of intestine gets stuck in the defect, the blood supply to the bowel may be blocked. This results in "strangulation" of that section of the intestine, which is extremely painful and needs immediate surgery so that the intestine can be freed and circulation restored. If the trapped intestine is damaged beyond repair, the affected part has to be cut out and the healthy parts joined together.

In the United States, the majority of hernia repair operations are carried out on an outpatient basis under local or general anaesthesia. A 1- to 2-inch incision is made over the site of the hernia and through the layers of muscle. The hernia is then pushed back through the defect in the abdominal wall, and the weak spot is repaired, sometimes using a mesh made of polypropylene or Gore-Tex that is stitched in place.

On average, the procedure takes 45–60 minutes; although if the procedure is done by "keyhole" surgery it takes longer. Painkillers are given after the operation to deal with pain around the site of the operation. Although a patient may resume normal activities within a week or two, heavy lifting must be avoided for three months to avoid straining the abdominal wall.

Some hernia operations are now performed by keyhole (laparoscopic) surgery, which allows a swifter recovery and less postoperative pain (see pages 128–129 for a step-by-step procedure of laparoscopic hernia repair).

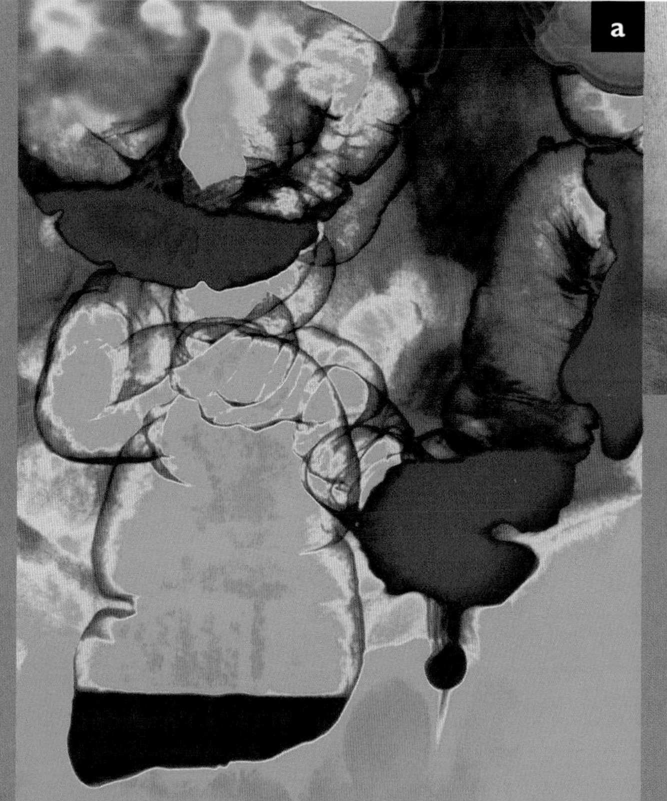

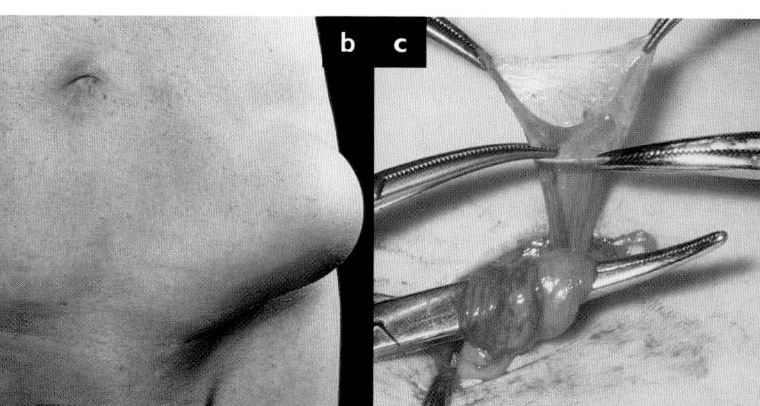

An inguinal hernia

a This contrast X-ray of an inguinal hernia shows barium pooled in the herniated part of the intestine (bottom left).

b A large hernia protrudes from the groin—it will need surgery to push the displaced intestine back inside the abdominal wall.

c After cutting through the tissue layers, surgeons locate the herniated section and repair the rupture.

Laparoscopic hernia repair

A relatively recent technique that was first used in 1982 but not widely introduced until 1990, laparoscopic hernia repair uses tiny incisions and delicate endoscopic instruments. Only certain surgeons choose to learn this technique, but patients can benefit considerably, with a quicker return to normal activity and less pain.

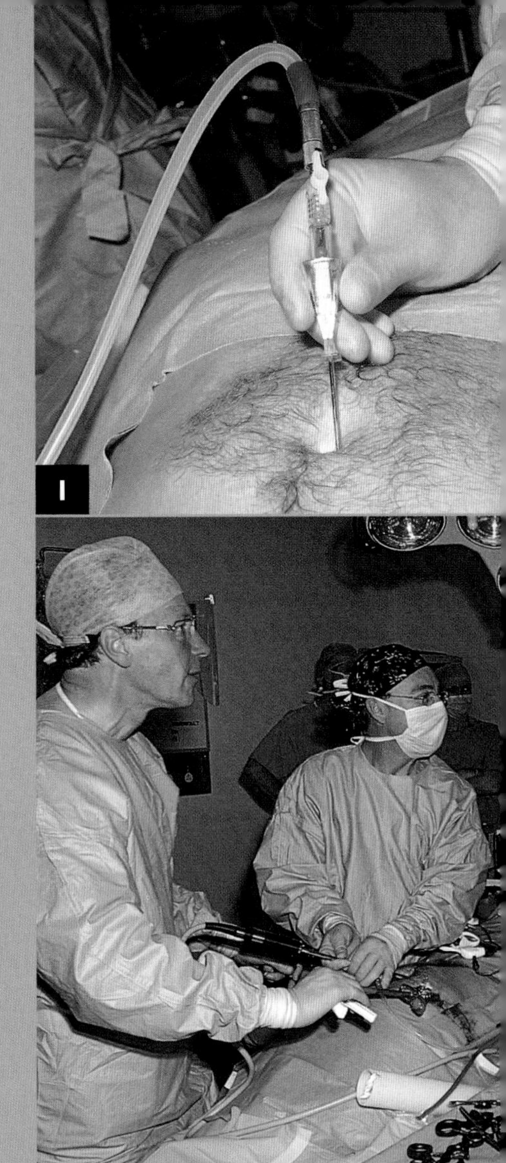

Hernia repair is one of the most commonly performed operations in the Western world—about 600,000 hernia repair operations are carried out each year in the United States alone. Until the 1990s, all these operations were performed by open surgery, which involves making a comparatively large—2-inch—opening in the abdominal wall. In contrast, laparoscopic hernia repair requires only tiny incisions through which the endoscopic instruments are passed. The technique, which is also known as minimally invasive or keyhole surgery, minimizes damage to the abdominal wall, so the patient heals more quickly, with smaller scars, and suffers less pain after the operation.

Open surgery remains the standard by which other techniques are judged, but laparoscopic hernia repair has become increasingly common in the last few years. Nevertheless, the procedure may not be offered by every hospital and clinic, for a number of reasons. Laparoscopic hernia repair requires extensive, specialized training, and its use is mainly restricted to the repair of inguinal and femoral hernias. Even for these conditions, there is still considerable debate over whether the new techniques are preferable to the standard ones (see page 127). The general consensus is that laparoscopic procedures are most suitable for treating recurrent hernias, and open surgery is often still best for treating first-time hernias. As research and expertise develop, however, the newer techniques may become more widely practiced.

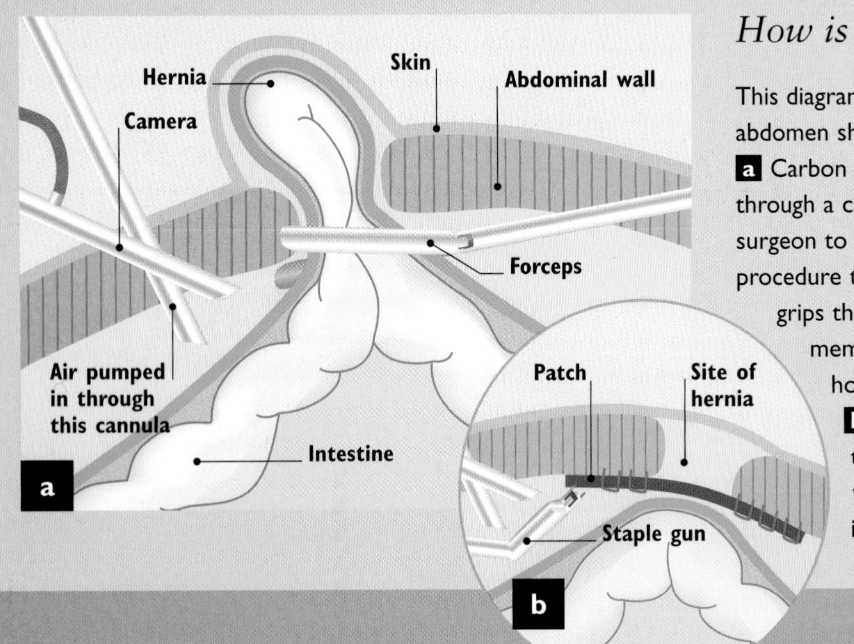

How is it done?

This diagrammatic cross section of a patient's abdomen shows the two main stages of surgery.

a Carbon dioxide is pumped into the abdominal space through a cannula (hollow tube), making space for the surgeon to work. Using forceps and viewing the procedure through an endoscopic camera, the surgeon grips the hernia (basically a loop of intestines in a membranous sac) and pulls it back through the hole in the abdominal wall.

b Once the hernia has been pulled back, the hole in the abdominal wall is patched up with a mesh patch, which is stapled to the inside of the wall.

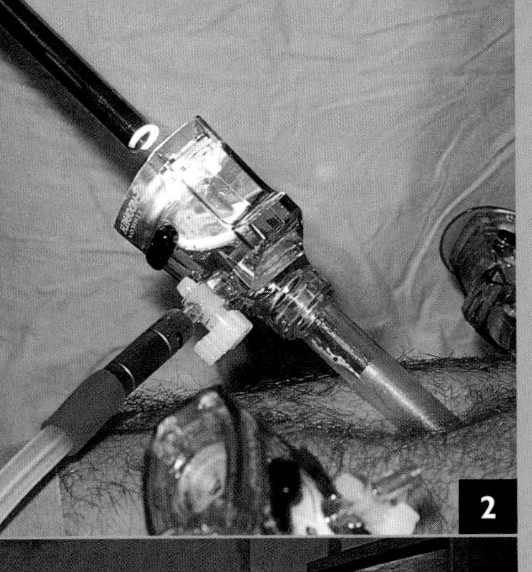
2

The procedure—step by step

1 Once the patient is fully anesthetized, and his abdomen swabbed with iodine, the surgeon makes three or four 1-inch incisions in the abdomen. In this case there are three—a central one near the belly button and one on either side.

2 Next, cannulas (hollow tubes down which instruments can be passed) are inserted through the incisions. Into one of these, the surgeon inserts a laparoscope, a telescopic video camera that allows him to see inside the abdominal cavity. Carbon dioxide is pumped through another cannula to make space to work in between the abdominal wall and the peritoneum (the membranous sac that contains the internal organs).

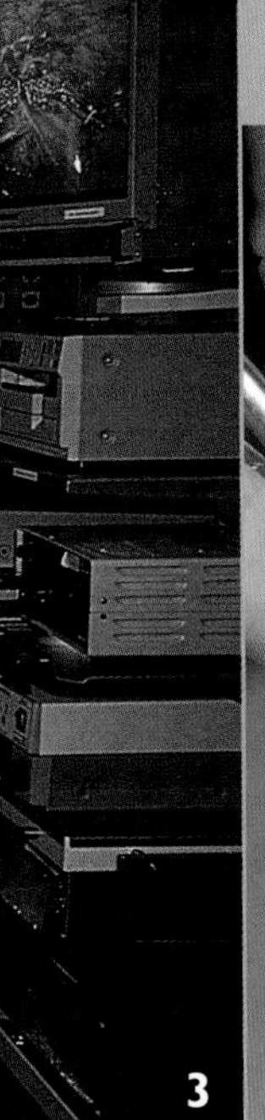
3

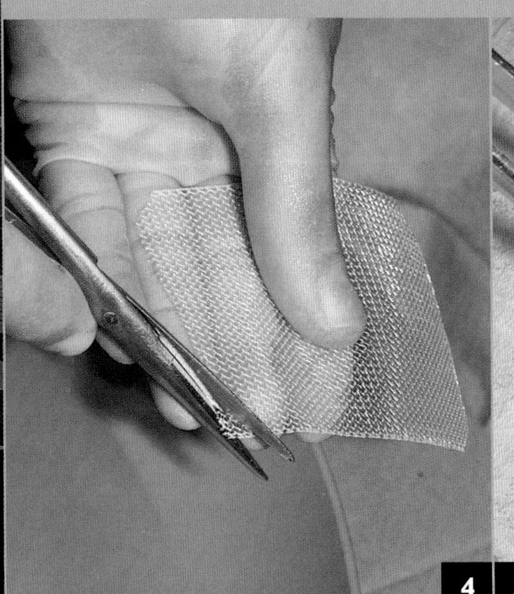
4

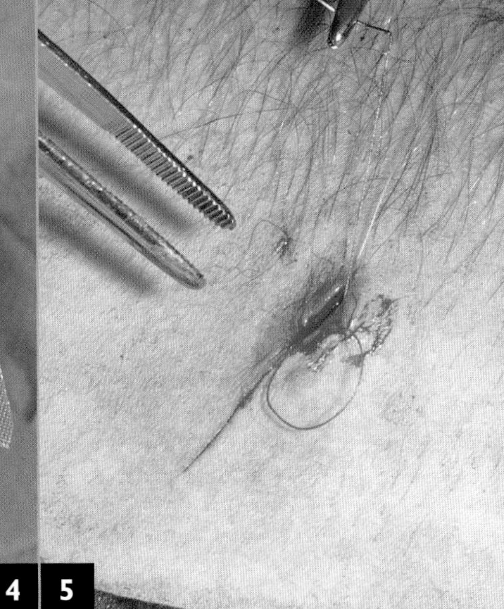
5

3 While an assistant steadies the cannulas, the surgeon passes endoscopic forceps and scissors down them, carefully watching the video screen to guide his movements. Using these instruments, he clears away connective tissue so that he can get a clear picture of what's going on. Once this is done, he grips the neck of the hernial sac with the forceps and pulls it back through the hole in the abdominal wall.

4 Now the hole needs to be patched. A piece of synthetic mesh is trimmed into shape, and then rolled up like a straw and fed down the central cannula. Once inside the patient, forceps inserted through cannulas on either side are used to unroll it and move it into position. A stapling device is introduced down one of the cannulas and the patch is stapled in place—surgeons use up to 30 staples. The abdominal space is deflated, and as the peritoneum returns to its normal position it squeezes up against the abdominal wall, helping to secure the patch in place.

5 The instruments and then the cannulas are removed and the tiny cuts are sewn up.

6 With the operation complete, the tiny wounds are dressed and the patient is taken to the recovery room. There is only minor scarring and the patient is often allowed to leave the hospital, in the company of a friend or relative, within six hours.

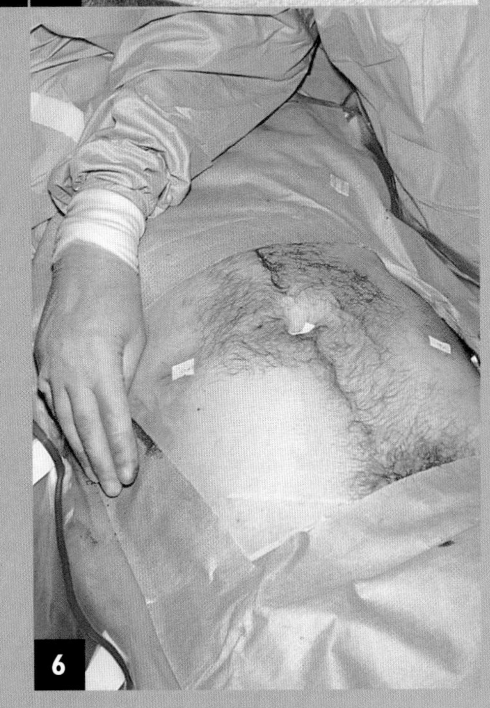
6

RELIEVING A GI OBSTRUCTION

An obstruction can occur anywhere along the digestive tract, from the esophagus to the rectum, bringing the digestive machinery to a standstill. The most common causes of obstruction are:

- strangulated hernias;
- adhesions following previous abdominal surgery;
- narrowing of the bowel due to inflammation, radiation, or infection; and
- tumors.

Without surgery, the part of the GI system nearest the obstruction stretches and becomes swollen and distended as it fills up with air and fluid. The GI system beyond the obstruction remains collapsed. The muscular wall of the GI system increases its contractions in an attempt to overcome the blockage. This causes colicky pain—intermittent, sharp, severe pain that occurs in waves—and constipation. Surgery is vital to remove the blockage and restore function to the GI system.

The average adult's small intestine is roughly four times longer than the person is tall.

Treatment for an obstructed esophagus

In some cases of inoperable cancer of the esophagus, the narrowed area can be stretched with a tiny balloon. Alternatively, a flexible metallic tube called a stent can be placed through the narrowed area to hold it open, allowing the individual to swallow. This may be enough to relieve symptoms. Where possible, an operation may be performed to remove the tumor. After esophageal surgery, patients are usually fed via a nasogastric tube.

Tumor removal

A tumor is a common cause of obstruction, but it may not become apparent until it blocks the tract completely or causes perforation. When surgeons remove a tumor, they aim to cut out a certain amount of healthy tissue around it to ensure that no cancerous tissue is left. The cut ends are joined together, if possible, to reestablish continuity of the gastrointestinal tract.

RE-ROUTING PARTS OF THE INTESTINE

The intestine may have to be re-routed surgically to bypass an obstruction or to rejoin two sections of the digestive tract after a diseased part has been removed. Ideally, the cut ends of the intestine are rejoined end to end, but if this not possible the intestine is brought up to the skin, where it is formed into a spout (made from the intestine itself) to empty into a plastic bag on the surface of the abdomen. Feces are then collected in the bag. The opening is called a stoma and may be permanent or temporary. The precise name of the operation depends on which part of intestine is brought to the skin. If it is the ileum (the last part of the small intestine) it is called an ileostomy; it is a colostomy if it's made from the colon. Both operations require a prolonged recovery period.

The procedure takes two to three hours and involves a week-long stay in the hospital. As well as the physical recovery, time is needed to learn about day-to-day stoma care and hygiene, and to adjust psychologically to having a stoma (see page 155).

Formation of an ileostomy

An ileostomy is usually formed on the right lower part of the abdomen. There are two types of ileostomy: end ileostomy, which is permanent, and loop ileostomy, a

Obstruction due to inflammatory bowel disease
This barium X-ray shows constriction and obstruction of a section of the intestine due to Crohn's disease (lower right). It may require surgery to remove the affected region and relieve the obstruction.

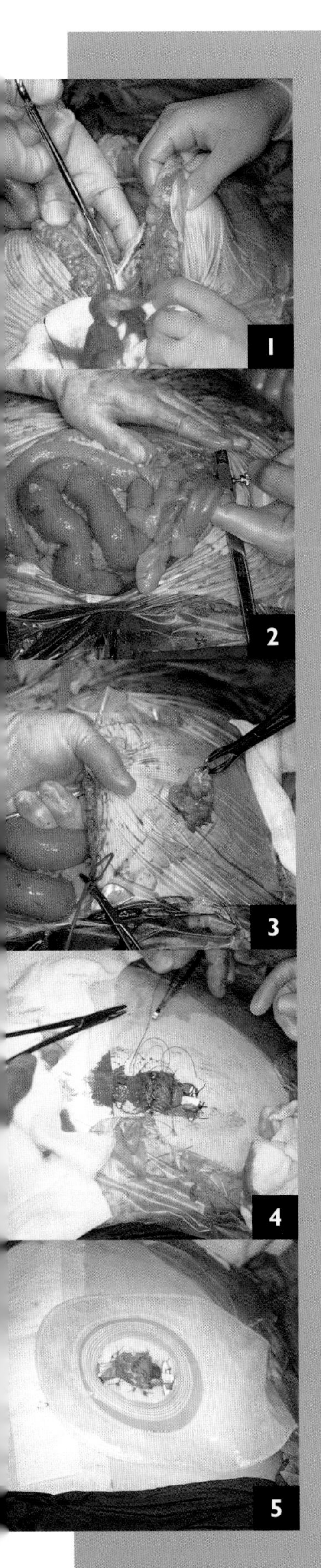

Step-by-step colostomy

1 The patient's abdomen is shaved, wiped with iodine, and covered in a clinging plastic wrap, which helps to guard against contamination. Surgeons cut directly through the wrap and the layers of skin, fat, and muscle that make up the abdominal wall.

2 Clamps are used to pull back and hold in position the layers of the abdominal wall, exposing the intestines. The diseased part of the colon is then cut out.

3 The upper cut end of the colon is brought to the surface through a small incision above and to the side of the belly button, separate from the main incision, so that a cuff of colon is left protruding. The lower cut end is sealed off with sutures.

4 The main incision is closed by suturing together the layers of the abdominal wall and the skin. The cuff of colon protruding from the smaller incision is folded back, and its outer edge is sewn to the skin to form the stoma.

5 The abdomen is cleaned, and the main incision is dressed. A colostomy bag is fitted over the stoma immediately after the surgery. It is either attached to a molded flange that adheres to the skin around the stoma, or, as shown here, attached directly via a sticky patch with a hole in the center, in the back of the bag.

Diversions in effect
This illustration shows the new route food will take on its journey through the digestive system after a colostomy operation. The colon is cut and brought to the surface to form a stoma (**1**)*, which will be connected to a colostomy bag. The damaged part of the colon* (**2**) *is bypassed.*

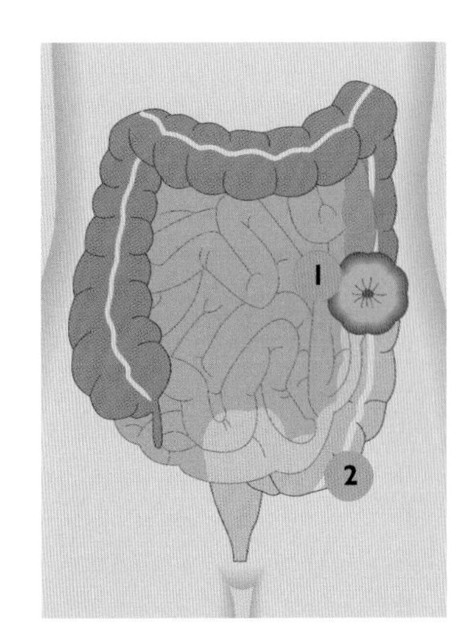

temporary, reversible procedure. An end ileostomy is usually formed if both colon and rectum have been removed, as may be necessary in cases of severe ulcerative colitis. The patient is taught how to empty, change, and dispose of the stoma bag, which needs to be worn at all times. Without proper care, the contents of the stoma bag may leak, causing skin irritation.

A loop ileostomy is formed to divert digested material away from the large intestine further along. A loop of ileum is brought up through the skin and opened in the middle at one side only. A spout is formed by turning the ileum inside out, allowing liquid feces to drain into a stoma bag attached to the skin. Feces that issue from an ileostomy are especially fluid, so care must be taken to avoid dehydration. A loop ileostomy can be reversed after several weeks by suturing or stapling the ends together.

Formation of a colostomy

A colostomy is the surgical creation of an opening of the colon onto the skin, similar to an ileostomy. A permanent end colostomy may be done when a tumor is removed and the two ends of the large intestine cannot be safely joined together. The end of the colon nearest the stomach is then brought through the abdominal wall. The end of the colon usually lies flush with the skin after the operation. The feces produced are fairly runny, as the colon is the part of the intestine that normally absorbs water from intestinal contents as they pass through.

A temporary loop colostomy is done to divert the intestinal contents away from the rest of the colon. A loop of colon is brought through the abdominal wall and kept in place with a glass or plastic rod. An opening is made in the loop of bowel to allow drainage into a colostomy bag. This can be reversed at a later date.

REMOVING PARTS OF THE INTESTINE

Several conditions can necessitate the removal of parts of the small or large intestine, including severe forms of inflammatory bowel disease (Crohn's disease and ulcerative colitis), and tumors of the intestine.

Crohn's disease often affects the small intestine, but it can affect the large intestine, too. In many cases, different areas of the intestine are affected simultaneously, with varying degrees of severity. Surgical removal of the diseased intestine is done as a last resort when a person's symptoms do not respond to drugs. Without surgery, persistent inflammation can cause severe narrowing of the intestine and lead to obstruction. In such cases, the narrowed part has to be removed. Crohn's disease can appear in any other part of the intestine, so surgeons usually remove only the smallest possible part. Between 30 and 50 percent of people with Crohn's disease have a recurrence of the problem after surgery.

In someone suffering from Crohn's colitis (Crohn's disease that specifically affects the colon), the colon can be "rested" by creating a temporary ileostomy or colostomy. In very severe cases, the whole of the colon may be removed and the person has a permanent stoma.

Banding hemorrhoids

a *First, the doctor places a short viewing device (proctoscope) inside the rectum. Then, the banding instrument is inserted through the proctoscope to grasp the hemorrhoid.* **b** *A rubber band is positioned around the swollen vein, which eventually shrinks and falls off.*

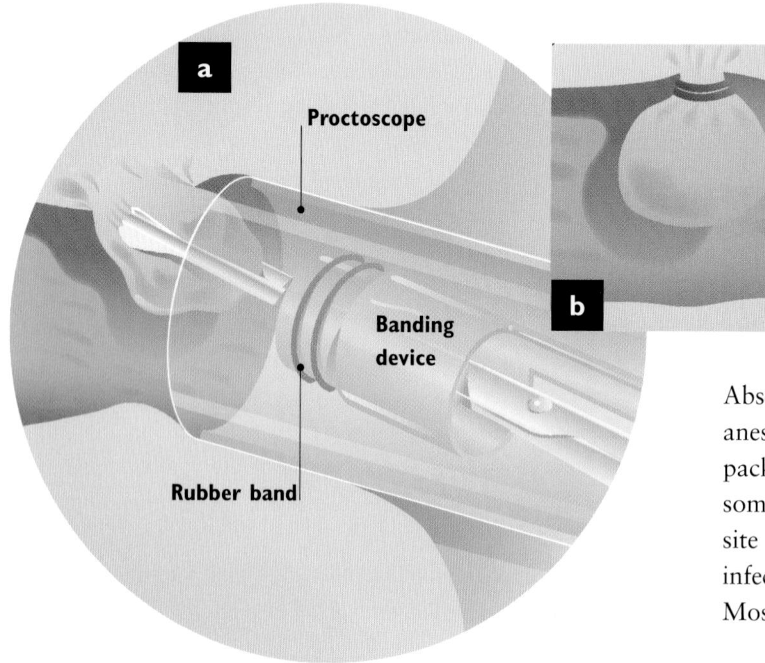

Colectomy—removing the colon

Removal of part or the whole of the colon is most often performed in individuals with ulcerative colitis. Long-standing ulcerative colitis can lead to changes in the lining of the colon and eventually may become cancerous. The risk of cancerous changes increases with severity and duration of the disease. Surgery is usually performed when the inflammation is severe and prolonged, resulting in a condition called toxic megacolon.

The colon is removed at least up to the rectum. When only part of the colon is removed, the end of the small intestine (the ileum) is brought out through the skin to form an ileostomy. In severe cases, the rectum is also removed. After this operation, the individual may have a permanent ileostomy. In some cases, however, a natural pouch may be formed from the ileum and attached to the anus so that feces can be passed normally. Patients tend to pass watery stools several times a day, but this gradually settles down as the intestine recuperates.

TREATMENT OF HEMORRHOIDS

Haemorrhoids (piles) are swollen veins in the anus. They usually cause bleeding from the back passage, an anal lump and discomfort. In many cases, hemorrhoids improve with a change of diet to relieve constipation and creams and suppositories to relieve inflammation and discomfort (see page 123). More severe hemorrhoids may need more radical treatment. One of the most common procedures for treating protruding hemorrhoids is band ligation (see left). Hemorrhoids that don't protrude from the anus can be treated by an injection of a chemical that seals the veins. In the most severe cases, piles can be removed surgically under general anesthesia.

REPAIR OF AN ANORECTAL ABSCESS

Anorectal abscesses—a collection of pus in or around the anus—are very common. Abscesses are removed and pus drained under general anesthesia. The cavity left behind is cleaned out and packed with gauze so that it heals from the inside. In some cases, an internal connection is found between the site of the abscess and the rectum, and through which an infection is perpetuated. This is called an anal fistula. Most fistulas are easily treated by surgery.

A TO Z
OF DISEASES AND DISORDERS

This section gives information on the main illnesses and medical conditions that can affect the digestive system. Liver and gallbladder disorders are covered in detail in another volume in the series.

This index is divided into two sections: Symptoms, and Diseases and Disorders. The entries are arranged alphabetically, and each is structured in a similar way:

What are the causes?

What are the symptoms?

How is it diagnosed?

What are the treatment options?

What is the outlook?

How can it be prevented or minimized?

SYMPTOMS

APPETITE LOSS

A common general symptom of illness rather than an indicator of a specific problem.

A loss of appetite may be related to nausea or abdominal pain in gastrointestinal disease, but in general it just represents the fact that an individual is unwell. Loss of appetite should be distinguished from difficulty in swallowing (dysphagia) and painful swallowing (odynophagia).

What are the causes?

Nausea, vomiting, abdominal bloating, or pain often result in a marked loss of appetite. In general, most illnesses will also, as mentioned earlier, result in loss of appetite. This is most marked in the presence of malignant tumors, which can cause weight loss and muscle wasting.

What are the treatment options?

Treatment depends on the cause of the appetite loss. Generally speaking, anything that makes the patient feel better also improves appetite.

CHANGES IN BOWEL HABIT

Any variation from normal in stool consistency and frequency of bowel movements

A change in the frequency and consistency of stools gives important clues about what is going on within the bowel. The presence of blood or mucus in stools is a strong indicator of abnormality.

What are the causes?

Basically, there are two changes: constipation or diarrhea. Constipation, characterized by a failure of the bowels to open, is often associated with hard stools. It can be brought on by a number of factors.

- Diet Lack of fiber in the diet and not drinking enough water are the most common causes of constipation.
- Age Deteriating muscle tone along the GI system can mean that stools take longer to pass through the system. Age is also often linked to a low-fiber diet.
- Medication Several drugs, such as codeine or aluminium-based antacids, can cause constipation.
- Medical conditions Underactivity of the thyroid gland and depression can cause constipation.
- An obstruction If the colon is narrowed or has tumors, it will impede the passage of stool.
- Anal pain Hemorrhoids or anal fissures can make passing stools painful and lead to a reluctance to open the bowels, eventually resulting in constipation.

Diarrhea involves increased frequency and volume of stools, which may be soft or watery. Mucus or blood may be present. Most causes fall into one of four main groups.

- Diet or infection Most short episodes of diarrhea are caused by food, drink, or a combination of the two that disagrees with the system, or to gastroenteritis.
- Medication Diarrhea is a side effect of several types of drugs, including antibiotics, ulcer-healing drugs, and senna.
- Medical conditions Overactivity of the thyroid gland (hyperthyroidism), diabetes mellitus, anxiety, and inflammatory bowel disease can all result in diarrhea. Celiac disease often results in loose, pale, bulky stools.
- Tumors Diarrhea, especially with bleeding or increased mucus in the stools, can be due to bowel tumors.

How is it diagnosed?

Doctors have strict definitions of what constitutes diarrhea or constipation, which may differ from a patient's own understanding of the terms. Nevertheless, patients reporting a persistent change in bowel habit, or

HELP YOUR DOCTOR TO HELP YOU

Monitoring bowel habit

A change in bowel habit can be an important symptom of several digestive problems. When taking your medical history, your doctor will want to know about the normal pattern of your bowel habits and about any changes to those patterns. Be prepared to answer the following questions as clearly as possible.

- *Before the onset of the symptoms that caused you to see the doctor, what was the normal frequency of your stools?*
- *Has the frequency, size, or hardness of your stools changed?*
- *Is there any blood in your stools? If so, how much and what color is it?*

Blood in the stools is a sign that you should visit your doctor immediately.

with rectal blood or mucus, need careful investigation to rule out a tumor. Initial investigations will include

Bowel habits depend on the person. A range from twice a day to twice a week all are entirely normal.

- examination of a stool sample for evidence of infection (p. 100);
- examination of the bowel itself, either with a barium enema (p. 108) or by sigmoidoscopy (p. 114) or colonoscopy (p. 115);
- small bowel barium follow-through (p. 107), if there is associated pain or a suspicion of celiac disease; and
- endoscopic biopsy (p. 112) of the duodenum if celiac disease is suspected.

What are the treatment options?

Apart from treating underlying causes, specific short-term measures can help to reduce the severity of bowel disorder symptoms. Drinking more water and using bulking agents that act as replacement dietary fiber can help with constipation. Drugs that increase secretion of fluid by the GI system, such as senna, are also useful, but must be taken only under medical supervision. Diarrhea can also be treated with drugs that slow down the muscle activity of the bowel wall, for example loperamide or codeine.

DIFFICULTY IN SWALLOWING

The sensation of food sticking in the throat or being difficult to swallow is known as dysphagia.

Difficulty in swallowing is a worrisome symptom for any patient. The difficulty may occur only when eating certain foods, for example bread or potatoes, or it may be so extreme that a person cannot swallow any solids or liquids. Dysphagia always requires careful investigation.

What are the causes?

Dysphagia may be caused by nerve or muscle problem in the esophagus that results in the failure of the swallow reflex; it is often seen after a stroke. The danger is that the food will "go down the wrong way" and end up in the lungs.

If food can be swallowed but then sticks, the cause is almost certainly in the esophagus. Irritation of the esophagus by stomach acid (esophagitis) is a common cause of mild symptoms, as are esophageal infections, such as *Candida*, for patients with weak immune systems. Severe symptoms indicate an obstruction—a physical blockage such as a tumor of the esophageal wall or pressure from swelling outside the esophagus.

Sudden-onset dysphagia may result from swallowing a solid object, such as a hard candy or a coin, or from a problem with the sphincter between the esophagus and the stomach, a condition known as achalasia.

How is it diagnosed?

The esophagus is examined with an endoscope, which gives a direct view of the wall of the esophagus and allows biopsies to be taken from it (p. 112). Sometimes a barium swallow is carried out to look for specific diseases such as achalasia.

What are the treatment options?

Patients with an obstruction of the esophagus may need to soften or liquidize food before eating. Narrowing of the esophagus can be treated by stretching with an endoscopic balloon or insertion of a stent (p. 124). Tumors can be treated with laser or conventional surgery, or radiation.

FATIGUE

A feeling of perpetual tiredness.

Fatigue is a symptom that many people report to their doctors and often represents the results of a "modern lifestyle" with the stresses of work and home life. Unless it is accompanied by other more specific symptoms, however, it can be difficult to work out the underlying problem. It is important to differentiate between tiredness attributable to overwork or lack of sleep and that caused by an underlying illness. It is also vital to be clear about the meaning of fatigue. Fatigue is often characterized by a perpetual feeling of tiredness, lethargy, and apathy; falling asleep without warning points to sleep-associated problems and represents an entirely different symptom.

Beyond the digestive system, fatigue can be caused by an underactive thyroid gland (hypothyroidism).

What are the causes?

Almost any illness will induce a sensation of tiredness, although a few specifics are important to consider.

- **Anemia** Low hemoglobin levels result in fatigue associated with shortness of breath, especially with mild exertion, and possibly ankle swelling. Other illnesses, particularly heart disease, can produce similar symptoms.
- **Inflammation of the bowel wall** Patients with Crohn's disease or ulcerative colitis frequently suffer fatigue, often in combination with diarrhea and abdominal pain.

- **Irritable bowel syndrome** Chronic fatigue is often associated with irritable bowel syndrome.
- **Malabsorption** If key nutrients are not absorbed, as in celiac diseases, fatigue results, often in combination with anemia.
- **Tumors in the bowel** As with any cancer, bowel tumors may cause tiredness along with weight loss and change in bowel habit.

How is it diagnosed?

The doctor's job is to find out if an underlying illness is causing the fatigue and in taking a medical history (p. 96) will try to identify appropriate next steps. A physical examination can help to diagnose hypothyroidism.

What are the treatment options?

Treatment depends on the causative illness. Fatigue associated with illness should resolve as the illness is treated and the patient's health improves.

How can it be prevented or minimized?

The fatigue of everyday life is common. To avoid becoming fatigued, follow these common-sense measures:

- Make time to eat properly.
- Find time to relax and exercise.
- Don't drink too much alcohol.
- Get enough sleep—most of us know how much sleep we can survive on and how much we need to feel good the next day; aim for the latter not the former.

JAUNDICE

Yellow discoloration of the skin and whites of the eyes.

Hemoglobin, the oxygen-carrying pigment in red blood cells, is broken down into a substance called bilirubin. The liver processes bilirubin into the bile salts that are excreted into the small intestine as bile, which helps in fat digestion and gives stools their brown color. A problem at any of the breakdown stages results in a build-up of bile salts, which are orange–yellow in color, in the tissues of the body—notably in the skin and the whites of the eyes.

What are the causes?

Three main problems result in jaundice.

- **Increased breakdown of red blood cells (hemolytic anemia)** In this condition, there is too much bilirubin for the liver to process; the overflow causes the yellow "staining."
- **Problems with the liver's ability to process bilirubin** Any illness involving the liver, such as hepatitis, alcohol abuse, inherited liver deficiencies, or liver tumors, will result in a build-up of bilirubin waiting to be processed.
- **Blocked bile duct** Gallstones, tumors of the pancreas, or rare tumors of the bile ducts can block the flow of bile from the liver to the small intestine. The processed bile salts spill back into the blood, causing jaundice. The failure of bile to reach the bowel results in pale waxy stools that look like putty.

How is it diagnosed?

After taking a careful history (p. 96), the doctor will probably use some other tests to narrow the diagnosis.

- **Blood tests** A sample of blood will be taken. A full blood count indicates whether there is any breakdown of red blood cells and reveals other abnormalities in the way the liver is functioning (p. 99). Blood analysis is also used to diagnose viral hepatitis and the causes of liver damage.
- **An ultrasound scan** The most common next step is a scan of the liver, gallbladder, and pancreas (p. 103). This test looks at the structure of the liver and reveals obstructions to the flow of bile or masses in the body of the liver.

Further tests will depend on what has been found.

- **ERCP** If an obstruction is suspected, an endoscopic retrograde cholangiopancreatogram (ERCP) is carried out (p. 111), often allowing an obstruction to be removed.
- **CT scanning** Computed tomography (CT) scanning is used to identify lesions within or obstructing masses outside the liver.
- **Biopsy** A liver biopsy, taken using a needle passed through the skin, can help pinpoint the cause of jaundice.

What are the treatment options?

Jaundice will disappear once the underlying cause is treated. Sometimes, drugs to alleviate associated symptoms, such as itchiness, may be prescribed.

NAUSEA AND VOMITING

Nausea is the sensation of needing to vomit. Vomiting is the forceful expulsion of the stomach contents through the mouth. Both are common symptoms.

Nausea and vomiting have very similar causes. The sensation of nausea and the physical action of vomiting are controlled by the "chemoreceptor trigger zone" (the CTZ, also known as the vomiting center), an area of the brain that receives information from the eye, inner ear, digestive tract, and bloodstream. Conflicting data from the eyes and inner ear can activate the CTZ, as happens with travel sickness and vertigo. The gut also has a strong sensory network of nerves that detect problems within the bowel and induce vomiting.

- **Gastroenteritis** This is possibly the most common cause of vomiting in adults. Infections can increase secretion of fluid in the digestive tract and at the same time slow muscle activity. The muscular walls of the GI system are stretched, producing a feeling of bloatedness. Vomiting is the body's way of relieving this discomfort.
- **Physical stimuli** Retching because of an object in the back of the throat or vomiting when the stomach becomes stretched, because of either overeating or failure of the stomach to empty, are both examples.
- **Toxins** Chemicals in your blood (such as poisons and chemotherapy agents) stimulate the CTZ directly, causing vomiting (often severe in the case of chemotherapy).
- **Emotion** Nausea and vomiting may be a response to unpleasant stimuli such as pain, fear, or anxiety, or the sight, sound, or smell of someone else being sick.

What are the treatment options?

Dealing with the cause of the vomiting is the best way of reducing nausea and vomiting. Drugs that reduce the symptoms (p. 116) act by blocking the initiation of vomiting or by helping the bowel to push food and fluid through the system, so that it cannot be regurgitated.

PAIN

Probably the most distressing of all symptoms, pain arises from anything that causes damage to the body, because of either a direct injury or inflammation.

Pain is detected by special pain receptors on specific types of nerves. The signal is then transmitted to the brain, which interprets the type and site of the pain. Although unpleasant, pain is a vital protection mechanism. In the digestive tract, pain causes a reduction in the amount of food eaten and encourages resting—both of which are important in allowing the system to rest and repair itself.

What are the causes?

Pain can be caused by damage to the digestive system from physical injury or trauma. It can also result from inflammation (as in appendicitis) or an obstruction of the bowel (as in Crohn's disease), which leads to stretching of the bowel wall. Strong contractions of the bowel (as in colic caused by gastroenteritis) are also painful.

What are the treatment options?

The primary goal is to treat the underlying disease or cause of the pain, but painkillers are important for relieving the symptoms. Simple analgesics, such as acetaminopen, are useful for abdominal pain caused by gastroenteritis. Strong opioid painkillers, such as codeine, are more effective but have a sedative effect and also cause constipation.

WEIGHT LOSS

Rapid loss of weight is often a warning of something more severe.

Rapid weight loss is one of the most worrying symptoms seen in a doctor's clinic. A patient with other symptoms whose weight is steady is unlikely to have a serious underlying illness.

What are the causes?

Pain or difficulty in eating may be caused by dental problems, such as ill-fitting dentures, ulcers, or abscesses. Pain or difficulty in swallowing usually stem from diseases of the esophagus, such as tumors, inflammation, infections (most commonly *Candida*), or a motility problem (such as achalasia).

Pernicious anemia damages cells in the stomach, resulting in severe B_{12} deficiency.

If there is no obvious cause at the "top end," then function is probably affected somewhere along the digestive tract. Failure to absorb nutrients from food (malabsorption), usually in association with bowel inflammation, is one cause of weight loss. Another is failure of the glands that play a role in digestion, as in chronic pancreatitis. Finally, weight loss may indicate a severe underlying illness, such as a tumor either within or outside the bowel.

How is it diagnosed?

Reaching a diagnosis involves careful assessment of all the patient's symptoms followed by investigations targeted toward suggested diagnoses.

What are the treatment options?

Treatment depends on the underlying cause of the weight loss, but ensuring an adequate calorie intake is the most important immediate treatment.

DISEASES AND DISORDERS

ACHALASIA

A problem with the sphincter at the end of the esophagus, which causes swallowing problems

Achalasia occurs in about 100,000 people in the United States each year.

In achalasia, the sphincter between the stomach and the esophagus will not relax, so food cannot pass into the stomach normally. The underlying cause is not known.

What are the symptoms?

Achalasia causes rapid onset of difficulty in swallowing. It can occur at any age. Food that is unable to pass into the stomach may regurgitate back up into the mouth.

How is it diagnosed?

A barium swallow (pp. 106–7) will show a baggy, dilated esophagus, with a narrow channel through the sphincter. Manometry (p. 102), to measure the pressure in the esophagus, or upper gastrointestinal endoscopy (p. 110) can help to confirm the diagnosis.

What are the treatment options?

Three different treatments are available:

- A balloon passed in via an endoscope can be inflated to stretch the sphincter, often producing a rapid improvement in symptoms.
- Small doses of botulinum toxin can be injected via an endoscope to relax the muscle in the sphincter.
- The outer layers of muscle around the sphincter can be cut to allow it to open.

What is the outlook?

Although the treatments are effective, only surgery is permanent—the others may have to be repeated.

AIDS, HIV, AND THE GI SYSTEM

Gastrointestinal illnesses that develop as a result of infection with the human immunodeficiency virus (HIV)

Infection with the human immunodeficiency virus (HIV) weakens the immune system, making it more vulnerable to infection. It can also lead to acquired immunodeficiency syndrome (AIDS), a combination of illnesses that includes regular and persistent gastrointestinal conditions.

What are the symptoms?

Most of the symptoms are caused by rare infections, known as opportunistic infections, which only tend to affect people with a compromised immune system; some arise because of rare cancers.

- **Difficulty chewing and swallowing because of pain** Fungal infections of the mouth and esophagus, particularly *Candida*, can cause painful inflammation in the mucus lining of the upper gut. Viruses such as herpes simplex and cytomegalovirus may also cause ulcers to develop.
- **Chronic diarrhea** Persistent and distressing diarrhea may occur because of a wide range of infectious organisms, including parasites such as *Cryptosporidium*, bacteria such as *Salmonella* and *Campylobacter* and some viruses.
- **Bowel tumors** Certain types of otherwise rare bowel cancers are seen in people suffering from AIDS. These include Kaposi's sarcoma and lymphoma. Such cancers can cause bleeding from the rectum and motility problems if they obstruct the bowel.
- **Jaundice** This "yellowing" of the skin may arise because of primary sclerosing cholangitis—a narrowing of the bile ducts between the liver and the bowel.

Specific conditions cause gastrointestinal symptoms such as nausea or diarrhea, but so do many of the drugs used to treat people infected with HIV.

How is it diagnosed?

The diagnostic tools used depend entirely on the symptoms. If diarrhea is a problem, doctors will analyse a stool specimen for parasites and bacteria. If a person is experiencing pain and difficulty swallowing, upper gastrointestinal endoscopy (p. 110) is usually carried out. Direct viewing of the bowel using a sigmoidoscopy (p. 114) or colonoscopy (p. 115) enables doctors to detect inflammation or ulceration. Many people with these symptoms have already been diagnosed as HIV positive. Occasionally, however, gastrointestinal symptoms caused by an unusual infection may be the first indication of a suppressed immune system. In these cases, doctors will advise the person to undergo HIV testing.

What are the treatment options?

The aim of treatment is twofold—to treat the particular condition causing the symptoms and, in addition, to use drug therapy to prevent replication of the virus and to improve immunity. Most infections can be effectively treated with appropriate antibiotics. If a cancer is causing the symptoms, chemotherapy is considered.

What is the outlook?

Advanced drug therapies now available to combat HIV infection mean that AIDS patients are much less likely to suffer from repeated gastrointestinal infections.

APPENDICITIS

Inflammation of the appendix

The appendix is a short, blind-ended tube that opens into the cecum. Inflammation and infection of this tube is known as acute appendicitis.

Acute appendicitis is the most common surgical emergency in the Western world.

What are the causes?

In many cases of appendicitis, the root cause is unclear, but in some cases inflammation arises because the appendix becomes blocked, usually by a fecalith—a small, hard ball of feces—which becomes lodged inside the central space. As a result, there is a build-up of secretions in the appendix and, if it is not removed, it becomes gangrenous and at risk of bursting and emptying its contents into the abdominal cavity. This can lead, in turn, to peritonitis.

What are the symptoms?

The exact location of the appendix varies from person to person, so the pain and symptoms associated with appendicitis are not always consistent. The classic symptoms, however, are fairly constant, and include

- pain that starts in the center of the abdomen, around the naval (belly button), and then moves to the right lower quarter of the abdomen;
- nausea and vomiting accompanying the pain;
- mild fever; and
- diarrhea.

How is it diagnosed?

The most helpful procedure in diagnosing appendicitis is a physical examination (p. 98). The doctor examines the abdomen to localize the painful area and will test for "rebound tenderness"—increased pain after removing pressure—in McBurney's point, a spot about halfway between the naval and the lower right corner of the abdomen. A rectal examination is done to try to pinpoint the source of pain, and then a sample of blood is taken to check the number of white blood cells. The presence of inflammation is confirmed if the levels are elevated.

What are the treatment options?

The only appropriate therapy for an acutely inflamed appendix is surgical removal (p. 126). Early surgery is advised in order to reduce the risk of peritonitis. The appendix is usually removed through a small incision in the lower abdomen or, less commonly, via a laparoscope passed through a tiny puncture site in the abdominal wall.

What is the outlook?

Most people recover rapidly from surgery for appendicitis, often being discharged about two days later. The scar heals quickly, and full recovery is the rule. If the appendix has ruptured, it can take longer to recover from the operation, as the bowel often stops functioning for a couple of days.

CANCER

See Tumors, p. 155.

CELIAC DISEASE

An inflammatory reaction to gluten, a dietary protein.

Gluten, a component of cereal crops such as wheat, rye, and barley, can cause an inflammatory reaction in the small intestine. This condition, known as celiac disease, results in the inability of the bowel to absorb certain nutrients, causing anemia, weight loss, abdominal discomfort, and diarrhea.

What is the cause?

Celiac disease is a direct result of hypersensitivity to gluten. Eating foods containing gluten results in damage to the wall of the small intestine. The usual lining of fingerlike projections, called villi, disappears, and the surface becomes flat. The loss of the villi reduces the surface area of the intestine so that it is unable to absorb nutrients from food. The underlying cause of the hypersensitivity is unknown.

What are the symptoms?

The symptoms of celiac disease can develop at any age but often first appear in people in their 20s and 30s. They include

- fatigue and generally feeling unwell;
- diarrhea and abdominal discomfort;
- mouth ulcers;
- muscle spasms or weakness; and
- a blistering rash on the arms and legs.

How is it diagnosed?

Experts use two types of test to make a diagnosis of celiac disease. A person with the condition will have specific types of antibodies in the blood, and so a sample of blood is taken for laboratory testing. In addition, an upper gastrointestinal endoscopy (p. 110) is used to take a small sample of tissue from the duodenum, the first part of the small intestine. When the tissue is examined under a microscope, the flattened appearance of the bowel caused by the loss of villi can be clearly seen.

Celiac disease tends to run in families and mainly affects people from Northern Europe. U.S. incidence is 1 in 4,700.

What are the treatment options?

Complete exclusion of gluten from the diet is the only way to allow the small intestine to repair itself. A single slice of bread is enough to cause a recurrence of the symptoms, so people with the condition are advised to adhere strictly to a gluten-free diet. The doctor or a nutritionist will give advice and guidelines on gluten-free foods.

What is the outlook?

At present there is no complete cure for celiac disease, although a gluten-free diet will enable the wall of the intestine to return to normal, allowing proper absorption of nutrients. Most people recover completely as long as they adhere strictly to a gluten-free diet.

CROHN'S DISEASE

See Inflammatory bowel disease, p. 148.

CYSTIC FIBROSIS

An inherited disease affecting the function of the pancreas, bowel, and lungs

Cystic fibrosis is a serious condition that results in recurrent chest infections and the failure of the pancreas to release enzymes that are vital for the breakdown and absorption of nutrients from food. Until recently, people with cystic fibrosis died in their teens, but advances in medical treatment mean this is no longer always the case.

What is the cause?

Cystic fibrosis is a genetically inherited condition—approximately 1 in 31 Americans unknowingly carries the gene abnormality, which is located on chromosome 7. It is a "recessive" disease, which means that a child with cystic fibrosis must have inherited a defective gene from both parents; if the gene is inherited from one parent only, the child has no symptoms but is a "carrier" of the disease and could pass it on to the next generation. Two parents who are cystic fibrosis carriers have a 1 in 4 chance of conceiving a child who suffers from the disease.

What are the symptoms?

The gene abnormality results in problems with secretion of mucus in the lungs, pancreas, and bowel, and it is this that causes the symptoms of the illness. These may include

- diarrhea and weight loss caused by a loss of pancreatic function—the pancreas cannot produce the enzymes essential for digestion of food;
- in babies, obstruction of the bowel by a very thick greenish bowel secretion (meconium); and
- recurrent chest infections with a persistent cough and breathlessness.

How is it diagnosed?

A family history of the disease is often a useful indication that cystic fibrosis is the underlying diagnosis. Initially, the doctor will take a blood test (p. 99) to look for the specific gene abnormalities that are present in cystic fibrosis. A "sweat test" may also be carried out. Children with cystic fibrosis have high levels of salt in their sweat—a sweat test is able to measure these levels.

Parents who are known to be carriers of the cystic fibrosis gene may be offered genetic counseling and fetal genetic screening when they are planning a family.

What are the treatment options?

There is a range of therapies offered to people suffering from cystic fibrosis, and individual treatment depends on the organs affected by the disease.

- **Pancreas** Artificial pancreatic supplements in the form of capsules can be given to replace the natural enzymes that are no longer being produced. The supplements are taken with meals to replace the patient's own pancreatic function and allow the normal digestion of food, relieving the diarrhea and malabsorption.
- **Gastrointestinal tract** In babies, blockage of the bowel by meconium requires surgical intervention to relieve the obstruction.
- **Lungs** The most important part of treating cystic fibrosis is therapy to keep the lungs as healthy as possible. It is advances in this field that have led to the great

ON THE CUTTING EDGE

Gene therapy for cystic fibrosis

Attempts are being made to use gene therapy in the treatment of cystic fibrosis. The aim is to replace the abnormal gene with a functioning copy. Harmless viruses are manipulated to carry the gene. The viruses are inhaled by the patient so that the gene enters cells in the lungs, allowing them to produce normal mucus. This can help patients avoid chest infections and reduce the damage to their lungs. As yet, the treatment is experimental and not entirely effective, but it is hoped that gene manipulation could one day offer real help to patients with diseases caused by gene abnormalities.

improvements in life expectancy. Early treatment with antibiotics for any sign of chest infection, coupled with regular, intensive physical therapy to keep the airways clear of mucus, is vital in order to avoid breathing problems.

What is the outlook?

Until ten years ago, children with cystic fibrosis died in their early teens because of overwhelming lung damage. Modern physical therapy and careful use of antibiotics has led to a dramatic improvement, with many sufferers now reaching adulthood with an improved quality of life.

DIVERTICULAR DISEASE

Pockets of bowel that protrude from the intestinal wall, which may bleed or become inflamed

The small pouches of intestinal lining (diverticula) that protrude from the intestinal wall in diverticular disease are common. Their exact cause is unclear, but it seems to be related to a Western diet that is low in fiber and high in protein. The condition is very common in Western countries but quite rare in the developing world, where vegetables form the staple diet.

What are the symptoms?

In themselves, diverticula are harmless, and 90 percent of affected people are completely unaware that they have diverticular disease. A proportion will suffer from mild diarrhea or constipation and mild pain in the lower left abdomen, which is relieved by passing wind or opening the bowels. However, complications can develop that cause the following symptoms.

- Constipation, a change in bowel habit, and rectal bleeding can occur in severe diverticular disease where strictures (narrowings) form in the part of the bowel most badly affected—usually the sigmoid colon.
- Severe abdominal pain, constipation, and fever can be a sign that a diverticulum has become inflamed—a condition known as diverticulitis. In severe attacks, abscesses may form, and if these rupture it can lead to peritonitis or obstruction of the bowel.
- Hemorrhage from blood vessels in the bowel wall adjacent to a diverticulum. The vessels may bleed profusely and cause shock from acute blood loss, although mild bleeding usually stops spontaneously.

How is it diagnosed?

In an otherwise healthy patient with only mild symptoms, the doctor may arrange a barium enema (p. 108) or a colonoscopy (p. 115). Diverticular disease gives the intestines a very distinctive appearance and is easily recognizable in both these tests. In people with severe diverticulitis, the diagnosis is made from the symptoms.

What are the treatment options?

In people with a mild form of diverticular disease, a high-fiber diet can help to avoid the development of complications, although treatment for constipation may be required in some cases. For diverticulitis and abscess formation, the doctor will prescribe antibiotics, intravenous fluids, and a period of fasting to allow the bowel to rest. Severe complications sometimes require emergency surgery to remove the affected part of the GI system.

DYSPEPSIA

Pain or discomfort in the upper part of the abdomen, commonly known as indigestion

Dyspepsia (from the Greek for "faulty digestion") is a broad term used to describe symptoms in the esophagus, stomach, and duodenum. These symptoms can be indicative of a serious problem such as peptic ulcers or *Helicobacter pylori* infection, which is now known to be a major cause of stomach ulcers. In most cases, however, there is no particular reason for the symptoms.

What are the symptoms?

Dyspepsia covers a wide range of symptoms that include the following:

- heartburn and acidity;
- belching;
- nausea;
- pain and abdominal distension; and
- an early feeling of fullness when eating.

Non-ulcer dyspepsia is a syndrome analogous to irritable bowel syndrome (p. 149). It is often stress-related and usually brought on by eating or drinking.

How is it diagnosed?

Upper intestinal endoscopy (p. 110) may be carried out to check for ulcers or a malignancy. When the symptoms are present but the stomach appears normal, non-ulcer dyspepsia is diagnosed.

What are the treatment options?

Most dyspepsia does not have a specific cause, so treatment involves medication to reduce discomfort, such as antispasmodics, antacids, and anti-gas agents. Non-ulcer dyspepsia sufferers should keep an eye on their diet and lifestyle to try to identify and avoid foods and behaviors, such as drinking and smoking, that cause problems.

FOOD ALLERGIES

An abnormal immune response triggered by a particular food

There are three different mechanisms that can give rise to gastrointestinal symptoms, all of which are often referred to as "food allergies":

- the normal response to irritants and toxins in food;
- an abnormal response to food because of an existing problem such as lactose intolerance; and
- a true allergic reaction to a specific food, such as peanuts or shellfish. A true food allergy occurs when the immune system is triggered by a particular food. This response involves the release of histamine, a chemical that makes the body react in an abnormal and sometimes life-threatening way.

What are the symptoms?

True food allergies will result in symptoms starting within an hour of the food being ingested. They include:

- swelling of the lips and tongue;
- vomiting;
- wheezing and breathlessness;
- eczema;
- migraine; and
- in severe cases, hypotension, tachycardia, and shock.

If the symptoms have delayed onset, making a direct connection with particular trigger foods is more complex. Abdominal discomfort with bloating and diarrhea are often blamed on food allergies but usually prove difficult to link with a single food item.

How is it diagnosed?

There are inherent difficulties in all allergy testing techniques, but the following may be used to help pinpoint the exact food that is causing the problem.

- **Skin patch testing** This involves placing a selection of possible allergens on the skin, then measuring any allergic response by the extent of redness or rash that develops. Often results do not match the foods the patient suspects.
- **Dietary exclusion** This involves systematically excluding particular foodstuffs from the diet to see whether symptoms improve. It can take a long time and often does not reveal useful results.
- **Dietary challenge** This technique involves feeding the patient a suspect food by itself to see whether it causes symptoms. Proper dietary challenge usually requires admission to a hospital and medical supervision.

HELP YOUR DOCTOR TO HELP YOU

Keeping a food diary

Food allergies that produce delayed reactions can be extremely hard to diagnose, because there is often no clear link between eating a trigger food and developing symptoms. One way to help your doctor identify your problem food is to keep a food diary—a record of food and drink, together with a list of your symptoms.

- *Make careful notes on what and when you eat and drink.*
- *Be sure to list all the ingredients of every meal—minor ingredients such as food additives are often allergic triggers.*
- *Record the timing, duration, and severity of your symptoms—use a scale to rate them.*

- **Food diary** This involves keeping a record of symptoms and foods, and trying to correlate the two (see box, p. 142). Some doctors do not consider food diaries to be scientifically valid, but many patients find them helpful.

What are the treatment options?

Simple exclusion of any trigger food is the most straightforward and effective way of treating true food allergies. Most patients have a good idea as to what causes their symptoms, particularly if they have kept a detailed food diary and are able to exclude the culprits from their diet.

The treatment of a life-threatening allergic response (anaphylaxis) is a medical emergency and requires injections of epinephrine, antihistamines, and steroids to halt the anaphylaxis and prevent severe breathing difficulties.

FOOD POISONING

See Gastroenteritis.

GASTRITIS

An inflammation of the wall of the stomach, resulting in mild symptoms

Gastritis simply means inflammation of the mucus lining of the stomach. It is a common condition that ranges in severity from being completely asymptomatic to causing extreme discomfort. When it does cause symptoms, gastritis results in upper abdominal pain that is sometimes made worse by eating, nausea, and a sensation of being full after only a small amount of food. Bloating, gas, and heartburn can also be associated with gastritis.

Helicobacter pylori is often contracted during childhood, but may not cause symptoms for many years.

What are the causes?

There are many causes of gastritis, including viral infection and alcohol abuse. Here, we look at the two main causes.

- **Nonsteroidal antiinflammatory drugs (NSAIDs)** This group of drugs, which includes aspirin, is used to treat a range of common conditions, from angina to arthritis. As a result, a large proportion of the population takes these drugs regularly. They can damage the protective lining of the stomach, resulting in inflammation.
- ***Helicobacter pylori*** This bacterium has been proven to be responsible for peptic ulcers and gastritis. It is responsible for most episodes of gastritis seen today.

How is it diagnosed?

Although the doctor may suspect gastritis from the symptoms (especially if a person has been taking an NSAID), the only way to obtain a definitive diagnosis is by performing an upper gastrointestinal endoscopy (p. 110), which allows the doctor to inspect the stomach lining directly. Blood tests, breath tests, and tissue sampling of the stomach lining may be performed and will confirm whether *Helicobacter pylori* is present.

What are the treatment options?

If a drug is the cause of the symptom, stopping it and changing to another type of medication is the only way of curing the gastritis. If the patient needs to stay on the medication, the doctor may prescribe a powerful drug called a proton pump inhibitor (p. 119) that turns off the acid production in the stomach. These drugs can reduce the risk of developing gastritis caused by taking NSAIDs. Infection with *Helicobacter pylori* is treated with a one-week prescription of two antibiotics plus a proton pump inhibitor, which successfully eradicates the infection in more than 90 percent of patients.

GASTROENTERITIS

Inflammation of the gastrointestinal tract

Of all gastrointestinal illness, gastroenteritis is the one that most of us will probably have suffered at least once. The condition is extremely common and is usually caused by a virus or bacterium in food that has not been destroyed during the cleaning or cooking process.

What are the causes?

A huge range of organisms can cause food poisoning. Many are viruses, such as rotavirus or astrovirus, which are especially common in children, or the Norwalk virus in older children and adults. Bacteria such as *Escherichia coli* (*E. coli*), *Salmonella*, *Campylobacter,* or *Shigella* are other major causes, as are parasitic microorganisms such as *Giardia*, amoebas, and *Cryptosporidium.*

What are the symptoms?

The familiar symptoms include nausea, vomiting, abdominal pain, and diarrhea. Sometimes, a person with gastroenteritis develops a high temperature. Severe infections with certain organisms, such as those that cause dysentery, can lead to bloody diarrhea. A very serious

Why do I get diarrhea on vacation?

Travelers' diarrhea is very common, affecting many people each year. As vacations are taken in more exotic locations, the number of attacks of gastroenteritis increases. The cause is often *E. coli*, sometimes *Shigella*, *Salmonella*, or a virus. It lasts for two to four days, with diarrhea and cramping abdominal pains. Good hygiene and careful selection of food at mealtimes provide the best way of avoiding an attack, because the infection is often caused by poor food preparation and the use of reheated food. If an attack jeopardizes a short vacation or business trip, a single dose of antibiotics (usually ciprofloxacin) can reduce the length and severity of the illness; otherwise, drinking plenty of fluids is the key to recovery.

ASK THE EXPERT

form of gastroenteritis is cholera, a bacterial infection that is now rare in developed countries; this causes torrential diarrhea and severe illness. Most attacks of gastroenteritis last for about two to four days and clear up without treatment. Affected people rarely visit the doctor, so the causative organism is not identified. If the illness persists, however, a visit to the doctor is essential, and a stool test can be arranged to identify the bacterium or parasite responsible for the infection.

What are the treatment options?

The most important part of treatment is maintaining good hydration by drinking plenty of fluids. Usually, plain water is enough, but in hot climates and in people with other illnesses, such as diabetes mellitus, maintaining the salt and water balance of the body's fluids with oral rehydration salts is recommended. These can be bought as small packets, or made using boiled water and orange juice with salt and sugar added. If a person has severe gastroenteritis, and diarrhea and vomiting are persistent, the doctor may need to give intravenous fluids. In some cases, antibiotics are given to shorten the length of the attack.

How can it be prevented or minimized?

Basic food hygiene is the key to avoiding gastroenteritis (pp. 43–45). Keeping cooked and raw meat separate, using different knives and chopping boards for meat and salads, and, above all, thoroughly washing hands and work surfaces before, during, and after food preparation, are vital in order to avoid infections.

GASTROINTESTINAL HEMORRHAGE

Bleeding from the bowel, which varies in severity

Any bleeding from the bowel is abnormal. However, it is not always serious and in many cases is caused by a minor ailment, such as hemorrhoids. Despite this, the bleeding should always be checked out by a doctor, because in some cases, it may be the first sign of a more serious condition. If bleeding is heavy, it can lead to anemia. If sudden and profuse blood loss is not treated promptly, shock can develop, which can be fatal.

What are the causes?

There are numerous causes for bleeding in the GI systems; peptic ulcers are the most common. In people with long-standing liver disease, the increased resistance of the liver to the flow of blood from the bowel results in dilated veins in the abdomen. In the esophagus, these veins are called varices and because they are under high pressure they blow up into balloonlike structures that bleed easily.

A condition known as angiodysplasia can cause bleeding from abnormal blood vessels in the wall of the lower bowel. Benign outgrowths from the bowel wall, called polyps, diverticula, and hemorrhoids, can all cause gastrointestinal hemorrhage, although the bleeding associated with these conditions is not usually heavy. When faced with bleeding from the lower GI system, a doctor's main concern will be to exclude cancer of the bowel. Malignancies usually bleed slowly, however, causing anemia rather than severe hemorrhage.

What are the symptoms?

The symptoms of bleeding will depend on where in the bowel the bleeding is originating. Blood loss from low in the bowel tends to be bright red in color, whereas if bleeding occurs in the stomach or upper intestine, the blood has gone through a digestive process and stools appear tarry and black.

Commonly experienced symptoms include

- vomiting, either of bright red blood or of partially digested blood (which resembles coffee grounds), caused by blood loss in the esophagus or stomach;
- smelly black stools known as melena;

- abdominal pain (notably when associated with an ulcer in the stomach); and
- nausea.

Often the gastrointestinal symptoms are accompanied by general symptoms of blood loss—dizziness, fainting, and shortness of breath.

How is it diagnosed?

The doctor will often be guided in making a diagnosis by the symptoms and type of bleeding. If the bleeding is causing vomiting or melena, then an upper gastrointestinal endoscopy (p. 110) is carried out so that the doctor can directly examine the esophagus and stomach for a bleeding point. Bleeding from the lower bowel is investigated using a colonoscopy (p. 115). If the site of blood loss remains unclear, an angiogram will be arranged. In this procedure, radioopaque dye is injected into the blood vessels of the GI system so that any bleeding into the bowel shows up on an X-ray. For people who have slow blood loss resulting in anemia, where doctors suspect that the bleeding is originating in the gastrointestinal tract, these same investigations will be carried out.

What are the treatment options?

With advances in technology, it is often possible to treat the source of bleeding during an endoscopic procedure (p. 125). Esophageal varices can be injected with a substance to seal them up, or a small rubber band can be placed around them to stop the flow of blood. Peptic ulcers can be injected with epinephrine or a sealant, or cauterized with a laser or heating device to stop them from bleeding. Bleeding in the intestine is treated in similar ways to lesions in the stomach.

With advances in endoscopic equipment, surgery is reserved for profuse bleeding that is not responding to any other treatment. Bleeding cancers, however, especially those in the colon or rectum, usually require the removal of the affected part of the colon (p. 132).

What is the outlook?

Although in many cases treatment is effective at stopping hemorrhage, bleeding from the upper part of the GI system can be an extremely serious, life-threatening condition. Recent data suggests that bleeding requiring emergency endoscopy may be fatal in 15 percent of patients; those at greatest risk are the elderly and patients with other serious medical problems. For others, it is the ailment causing the bleeding that determines its outcome.

GASTROESOPHAGEAL REFLUX

The movement of the contents of the stomach back up into the esophagus

Gastroesophageal reflux disorder (GERD) is extremely common. In fact, reflux is a normal occurrence and it is only when symptoms appear that it is labeled a medical condition. These symptoms are caused by stomach acid leaking back up from the stomach into the esophagus, irritating the more sensitive lining of the esophagus and causing pain. In the majority of cases, the valve or sphincter at the entrance to the stomach appears normal, although some patients will have an anatomical abnormality, such as a hiatal hernia, that disrupts the function of the valve.

What are the symptoms?

In the majority of people, the main symptom is heartburn—a burning pain running up and down the center of the chest. The pain is often brought on by particular activities, such as bending over when gardening, or by certain foods or alcohol. Some people experience stomach acid or bile entering the back of the mouth, causing a bitter taste. In addition to these symptoms, there are three main complications of GERD.

Some 40 percent of the population suffer from heartburn on a monthly basis.

- **Esophagitis** Acid damage can cause inflammation and even ulcers in the lining of the esophagus.
- **Strictures** In severe, long-term GERD, the esophagus can become narrowed.
- **Barrett's esophagus** This occurs when the cells in the lining of the esophagus undergo a change and resemble those in the stomach or small intestine. The cause of this change is not clear, although it is almost certainly a protective response to the stomach fluids attacking the esophagus. Barrett's esophagus can increase the risk of developing cancer so a program of regular endoscopy may be recommended to pick up early signs of tumors.

How is it diagnosed?

The doctor usually makes a diagnosis of GERD based on the classic symptoms. For a young person with no other symptoms, the doctor is likely to advise lifestyle changes and medication. In people over the age of 45 and those who have other symptoms, upper gastrointestinal endoscopy (p. 110) is carried out in order to eliminate other conditions, such as a hiatal hernia, and to determine the extent of damage to the esophagus. Studies to measure the acidity in the esophagus may also be done (p. 102).

HELP YOUR DOCTOR TO HELP YOU

Helping with heartburn

Heartburn is a very common problem. One in every 20 patients seen by a physician is there because of heartburn or indigestion. It is the doctor's job to rule out any other, more serious causes for the symptoms, and this involves looking for signs of a significant problem in the esophagus. The doctor will want to know if you are experiencing any of the following:

- *weight loss;*
- *difficulty or pain in swallowing; and*
- *vomiting, especially if there is blood in the vomit.*

If these symptoms are absent, it is less likely that there is a serious disease causing the problem.

What are the treatment options?

Treatment focuses on reducing the amount of acid in the stomach and preventing it from reaching the esophagus. Certain changes in lifestyle, such as reducing alcohol intake and eating fewer rich foods, can help to relieve symptoms. In severe cases, some people may benefit by raising the head of the bed so that they do not lie flat at night.

Neutralizing the refluxing acid during an attack should resolve the symptoms quickly. Effective antacids are available over the counter, either as chewable tablets or as a liquid, and should be taken early during an attack. Doctors may also prescribe preventive medicines, such as ranitidine and omeprazole, which act by reducing the production of acid in the stomach (p. 119).

In severe cases, hiatal hernias may be repaired surgically. Laparoscopy-assisted surgery can improve the efficiency of the valve between the esophagus and the stomach.

What is the outlook?

GERD responds very well to antacid drug treatment. Strictures are rare these days, but even if they do develop, the narrowed area can be dilated during endoscopy. Where surgery is necessary, the outcome is generally excellent, although after the operation some people suffer from the sensation of food sticking in the gullet.

How can it be prevented or minimized?

Lifestyle changes can reduce both the severity and the frequency of symptoms—rich foods, alcohol, and smoking all contribute to GERD, as does being overweight. Even minimal weight loss can help to prevent the reflux of stomach acid into the esophagus.

HEMORRHOIDS

Dilated veins in the rectum that cause soreness and bleeding

Hemorrhoids, commonly known as piles, are dilated veins just inside the rectum. They are associated with the presence of small tags of skin around the anus and often protrude through the anus, causing irritation. Hemorrhoids frequently develop in association with constipation and having to strain in order to pass stools; they may also occur during pregnancy. If hemorrhoids are long-standing, the blood inside them clots, making them hard and sometimes painful.

What are the symptoms?

Not everyone with hemorrhoids has symptoms, but for those who do, the main complaint is itching. The presence of small amounts of bright red blood on the surface of stools or on the toilet paper after use is another common sign. If the hemorrhoids become large enough to protrude through the anus—a complication associated with straining to open the bowels—they can become extremely painful.

How are they diagnosed?

Diagnosis is usually made with proctoscopy, in which a short, hollow viewing instrument is inserted through the anus and slowly withdrawn. Any hemorrhoids are clearly visible as the tube is pulled out. If the doctor is concerned about bleeding from the rectum, an examination using a flexible endoscope (p. 114) may be carried out. This is because bleeding from the lower bowel can arise from several more serious diseases, including cancer.

What are the treatment options?

Basic changes in lifestyle, such as increasing dietary fiber and fluid intake to avoid constipation and straining, will help to reduce the symptoms. If the hemorrhoids are large or painful, the doctor may inject them with a sclerosing agent that causes them to shrivel up. In some cases, surgery is required to remove the veins completely (p. 132).

HEARTBURN

See Gastroesophageal reflux, p. 145.

HERNIA

An abnormal protrusion of intestine through a gap in a layer of tissue

Hernias occur when the pressure within the abdomen pushes a part of the intestine through a weakened area in the muscle wall so that it protrudes and can be felt as a lump. A hernia can occur in the groin or scrotum (inguinal), where the leg joins the abdomen (femoral), or through the bellybutton (umbilical). Hiatal hernias affect the stomach and esophagus (see below).

What are the symptoms?

The soft mass of intestine is often clearly visible as an uncomfortable or painful swelling. In severe situations, the part of intestine in the hernia can twist, cutting off its blood supply and causing acute pain. If this condition, known as a strangulated hernia, remains untreated, the affected tissue can die from a lack of blood supply. Strangulated hernias are most commonly seen in the femoral area and require urgent surgery.

How is it diagnosed?

A hernia can be easily diagnosed from the physical symptoms it causes, including the appearance of a lump and a dragging or aching sensation.

What are the treatment options?

Surgeons usually advise that hernias be repaired to avoid strangulation. Surgical repair, which can be done either by open surgery (p. 127) or by laparoscopic surgery (p. 128), involves pushing the intestine back into the abdomen and then reinforcing the weak area in the abdominal wall. Although the surgery is straightforward, hernias sometimes recur. If there is a delay before surgery, a truss can be worn to support the weak area.

HIATAL HERNIA

Part of the stomach protrudes into the chest cavity

The gap in the diaphragm where the esophagus passes through on its way to the stomach is called the hiatus. If the diaphragm becomes weakened, most commonly because of age, pressure from the abdomen can force part of the stomach through this gap, forming a hiatal hernia. Normally the hiatus acts like a sphincter, helping to control the opening between the stomach and the esophagus and preventing stomach acid from spilling into the esophagus. In a hiatal hernia, the sphincter can no longer function properly and stomach acid can leak back (reflux).

What are the symptoms?

Acid reflux produces the symptoms of gastroesophageal reflux (heartburn) (p. 145), especially when lying flat. The hernia itself can interfere with swallowing, giving the sensation of food sticking in the throat.

How is it diagnosed?

Hiatal hernia is diagnosed with an upper gastrointestinal endoscopy (p. 110) or a barium swallow (pp. 106–7).

What are the treatment options?

Symptoms of heartburn can generally be controlled with antacids and other medication, or with lifestyle measures such as a healthy diet, losing weight, giving up smoking, and raising the head of the bed at night. In severe cases, surgery can repair the hernia, but many recur.

EXPERIENCING AN INGUINAL HERNIA

I first noticed the bump just after doing some work in the garden. It was about 2 inches across, in my groin just above my testicle. It didn't hurt, and would come and go.

Gradually it got bigger, although I could easily push it back in. Eventually the testicle became enlarged, and I'd get a dragging sensation in it, so I went to see my doctor who diagnosed an inguinal hernia and suggested I wear a truss to ease the discomfort. He referred me to a surgeon in the local hospital, who put me on the waiting list to have it repaired.

After the operation, I was a little sore. I was allowed to go home the next day on strict orders not to do any gardening or lift anything for six weeks. Now I feel as good as new, and the only sign of my hernia is a 4-inch scar, hidden in the crease of my groin.

HIRSCHSPRUNG'S DISEASE

A congenital disease that appears in early childhood, causing constipation and a dilated colon.

Hirschsprung's disease is a rare congenital condition that affects about 1 in 5,000 babies. The symptoms are caused by a malformation of the nerves that control the movement of the intestine. In a healthy person, waves of contractions (peristalsis) run along the intestinal walls, pushing food through the gut, rather like pushing toothpaste out of a tube. The signals that propagate the muscle contractions are transmitted via the nerves of the intestinal wall. In a child with Hirschsprung's disease, part of the nerve system is missing, so peristalsis stops at that point and the bowel becomes obstructed.

Hirschsprung's disease is four times more common in boys than in girls.

What are the symptoms?

The doctor will suspect Hirschsprung's disease if a child has severe constipation or an obstructed bowel in the first couple of years of life, although if the abnormality is minor, the condition may not cause problems until adulthood. An affected child will have the following symptoms:

- difficulty emptying the bowels;
- colicky pain;
- a distended abdomen;
- poor appetite; and
- vomiting, in severe cases.

How is it diagnosed?

Usually, the first investigation is a barium enema (p. 108). In a child with Hirschsprung's disease, the colon will be much wider than normal. The diagnosis is confirmed by measuring the pressure of the sphincter at the anus, which fails to relax, or by taking a biopsy (a tissue sample). This can be done during surgery or during a proctoscopy procedure (p. 114). Microscopic examination of the biopsy tissue will show whether nerves are absent from the affected part of the large intestine.

What are the treatment options?

Surgery to remove the part of the bowel that has no nerve supply, and to rejoin the unaffected areas, is usually curative, with no long-term complications. A temporary colostomy may be necessary before the main surgery, if the child is seriously underweight and needs urgent restoration of at least partial GI function.

INDIGESTION

See Dyspepsia, p. 141.

INFLAMMATORY BOWEL DISEASE

A term that describes two key diseases in which the bowel becomes inflamed—Crohn's disease and ulcerative colitis

Crohn's disease and ulcerative colitis overlap considerably in their signs and symptoms, and doctors categorize them both under the title of inflammatory bowel disease. One difference, however, is that Crohn's disease can affect any part of the bowel, whereas ulcerative colitis is confined to the colon and rectum.

What are the causes?

There are many theories as to the cause of these diseases, but none have been fully proven. Certain infections (the measles virus, live vaccines, and mycobacteria) have all been proposed as being responsible, but the evidence is not strong enough to pin the blame on any of these with certainty. In both diseases, the body's immune system seems to become abnormally stimulated, and this becomes self-perpetuating, resulting in inflammation and ulceration of the wall of the intestine. It is this inflammation that causes damage to the lining of the gut and the resulting symptoms.

What are the symptoms?

Although the symptoms of these two conditions are often so similar that doctors find it difficult to tell them apart, there are a few distinguishing features.

- **Ulcerative colitis** Diarrhea is the main symptom, with bowel movements occurring up to 15 to 20 times a day when an attack is underway, often with blood and mucus in the stool. The loss of blood may result in anemia. In a very severe attack, the colon may become dilated, a dangerous complication that results in fever, feeling very unwell, and abdominal pain with the diarrhea.
- **Crohn's disease** Because it can involve the whole of the digestive tract, from mouth to anus, Crohn's disease can cause a huge array of symptoms. Mouth ulcers are often a problem. In the small intestine, inflammation results in pain, narrowing, and even obstruction of the bowel, leading to vomiting. The ileum (the final part of the small intestine) is commonly affected, causing pain, obstruction, and sometimes problems absorbing vitamin B_{12}. Crohn's disease in the colon can be similar to ulcerative colitis,

with the passage of bloody diarrhea caused by inflammation, and formation of ulcers. Passages, called fistulae, can form between the bowel and the skin.

How is it diagnosed?

The doctor will probably suspect inflammatory bowel disease from the symptoms; blood tests will be done to look for indications of inflammation and also to check for anemia (p. 99). The diagnosis is then confirmed by taking a small sample of tissue—a biopsy—from the intestinal wall during colonoscopy (p. 115). If the doctor suspects that Crohn's disease has affected the small intestine, a small bowel follow-through (pp. 106–7) usually highlights any areas of inflammation or narrowing.

What are the treatment options?

There are four main treatment options.

- **Antiinflammatory drugs** These drugs, for example mesalamin, are used to reduce the inflammation in the bowel wall. It is important not to confuse these with antiinflammatories of the aspirin family, which can make the symptoms worse.
- **Immunosuppressant drugs** These are used to reduce the immune response in the bowel. Steroids can be useful, but side effects may limit the dose. Drugs such as azathioprine help to reduce the level of steroids needed.
- **Surgery** Generally, doctors prefer to avoid surgery, but in patients with severe symptoms that are not responding to medication, the affected part of the bowel may need to be surgically removed.

ON THE CUTTING EDGE

Antibody therapy in Crohn's disease

Recently, a new type of treatment for Crohn's has become available. The inflammation in Crohn's disease is at least partially caused by a chemical the body produces, called tissue necrosis factor (TNF). TNF attaches to cells in the intestine at a specific part of the cell wall (a receptor) and in doing so triggers the inflammation. Antibody therapy aims to prevent the attachment of TNF to the cell wall by binding to TNF, breaking the cycle of inflammation. The antibodies are administered as a drug, infliximab, which has been shown to be helpful in certain types of Crohn's disease, stopping attacks that did not respond to other treatments.

- **Diet** A low-fiber, high-vitamin diet may help to alleviate symptoms, and some patients also claim that avoiding certain "trigger" foods produces improvements.

What is the outlook?

Both Crohn's and ulcerative colitis are "relapsing and remitting" diseases. In other words, they flare up from time to time, often without warning. Although medication can reduce the frequency and severity of attacks, it does not cure them altogether. Because ulcerative colitis affects only the colon, removal of the colon effectively cures the disease, but may leave the patient with a stoma, or bag, and is usually reserved for severe cases (p. 131). Crohn's can affect any part of the digestive tract, so surgery to remove one portion may not cure the disease. In fact, repeated surgery can result in a marked reduction in the length of the bowel and consequent difficulty absorbing nutrients.

IRRITABLE BOWEL SYNDROME

A collection of gastrointestinal symptoms that are thought to be stress related.

Irritable bowel syndrome (IBS), also known as functional bowel disorder, is extremely common and affects women more than men. The syndrome is a collection of symptoms that arise in the GI system. In its mildest form, it develops at times of stress, such as before exams or interviews, and causes diarrhea. In some people, however, the condition becomes continuous and leads to more upsetting symptoms.

What are the causes?

As yet we do not know why some people suffer severe symptoms and some do not. Stress can bring on attacks or make them worse for some people, whereas for others certain foods are the trigger—fruit and the artificial sweetener sorbitol are common culprits. The anatomy of the intestines appears completely normal in people with IBS, although recently abnormalities in some of the chemicals in the intestinal wall have been found in some sufferers.

What are the symptoms?

The symptoms of irritable bowel syndrome are numerous and vary among individuals. A person with the syndrome may have one or more of the following:

- indigestion (dyspepsia) and burping;
- gastroesophageal reflux (heartburn);
- nausea and fatigue;

EXPERIENCING IRRITABLE BOWEL SYNDROME

My first "attack" was in my late 20s. It seemed I'd inherited the "sensitive tummy" of my family; my dad and two aunts suffered terribly with digestive problems.

I thought it was food poisoning at first, but the diarrhea would come and go, and the pain was intense. Then I considered a food intolerance and kept a food diary for months, but I never identified a culprit. After another severe attack, I went to see the doctor again.

My doctor asked about my lifestyle and my job, which was going through a stressful time, and suggested that I had irritable bowel syndrome. The abdominal pain was due to rapid contractions of my intestines, so she prescribed an antispasmodic drug. Since then I've learned to handle stress better and generally the IBS is under control, but I still suffer symptoms occasionally.

- pain in the upper abdomen or, if the colon is the main site, colicky abdominal pain; and
- constipation, diarrhea, and bloating.

The sensation of needing to open the bowels again immediately after going to the toilet is common, along with attacks of diarrhea in the morning, which then subside for the rest of the day.

How is it diagnosed?

Before making a diagnosis of IBS, it is important to rule out other gastrointestinal conditions that cause discomfort and a change in bowel habit. If there is any doubt about the diagnosis, a colonoscopy (p. 115) or barium enema (p. 108) is often carried out.

What are the treatment options?

Identifying and avoiding trigger factors and adopting a healthy lifestyle are the most important general measures, but no single treatment will work for everybody with IBS, and there is no apparent cure for all the symptoms. Several medications may be tried until individual symptoms are relieved. The following treatments may help:

- Peppermint oil capsules or dimethicone can help bloating and burping.
- Antispasmodics, such as hyoscyamine, glycopyrrolate, and dicyclomine may relieve abdominal pain.
- Antidiarrheals, such as loperamide, are sometimes used to treat severe diarrhea.
- High-fiber diets often improve symptoms in those with marked constipation, but can exacerbate bloating.
- In a few people, drugs are used to modify the way the disturbances in the bowel are perceived. This involves the use of drugs that were originally designed as anti-depressants, such as prothiaden. Although not everyone with IBS is depressed, evidence suggests that these drugs can markedly improve symptoms, especially if pain is a major symptom.

Hypnotherapy and psychotherapy have been shown to be helpful in many patients with IBS, with sessions targeted at improving GI system symptoms. Such approaches are becoming more widespread and more easily available.

ISCHEMIC BOWEL DISEASE

Loss of the blood supply to the intestines, resulting in a range of symptoms

Bowel ischemia can be described as a "heart attack of the gut." Just as a heart attack is caused by blockage of the blood vessels supplying blood to heart muscle, so bowel ischemia is due to blockage of the arteries in the bowel.

What are the causes?

Atherosclerosis, in which patches of hard cholesterol form in the walls of the arteries, is the main cause of bowel ischemia. These "plaques" gradually build up until the artery becomes narrowed or blocked, reducing the supply of oxygenated blood to the bowel. Sometimes a blood clot travels from another part of the body and blocks the artery. Smoking, high blood pressure, diabetes, and high cholesterol all increase the risk of bowel ischemia.

What are the symptoms?

The symptoms of bowel ischemia vary depending on the extent of the blockage. Mild ischemia results in gripping, sometimes severe, abdominal pain, notably after eating, when the blood flow requirements of the system increase but the narrowed arteries cannot supply the demand.

If the main artery to the bowel, the mesenteric artery, becomes blocked, the gastrointestinal equivalent of a heart attack occurs. The affected bowel becomes starved of blood and oxygen and without immediate treatment, the affected part can be severely damaged, or even die, leading to perforation and peritonitis. If this occurs, the patient suffers severe abdominal pain and vomiting. Bloody diarrhea occurs as the attack progresses.

How is it diagnosed?

If the doctor suspects a gradual narrowing of the arteries, an angiogram will probably be arranged. In this procedure, radioopaque dye is injected into the main artery of the bowel, via a catheter in the groin, and a series of X-rays is taken that will show up any narrowed areas. Acute bowel ischemia usually causes symptoms severe enough to require emergency exploratory surgery. The appearance of the bowel during the operation confirms the diagnosis.

What are the treatment options?

Narrowing of the arteries can be treated by threading a catheter into the affected vessel and inflating a tiny balloon in the narrowed segment. In some cases, drugs help to improve the flow of blood in the bowel.

Complete obstruction of the blood flow to the bowel is treated by surgically removing the dying section of bowel. The outlook for surgical patients is good, especially if treatment is prompt and the bowel has not perforated. Often, however, nutritional problems will result following the removal of a large section of bowel.

LACTOSE INTOLERANCE

Gastrointestinal symptoms related to the consumption of dairy products

Some people are unable to consume dairy products without experiencing abdominal discomfort and bloating along with nausea and diarrhea two to three hours afterward. These people are described as having lactose intolerance—a situation in which the GI system is unable to break down the sugar in cow's milk, known as lactose.

What are the causes?

In order for lactose to be digested and absorbed from the GI system, it must be broken down into its two constituent sugars, glucose and galactose. The enzyme responsible for this breakdown process is lactase, which is found in the cells lining the small intestine.

For adults in the majority of the world's ethnic groups, lactose intolerance is the norm. The ability to produce lactase is lost at weaning, but this does not generally cause problems because dairy products are largely absent from the diet in such cultures. In the Western world, however, where dairy products feature heavily in the normal diet and populations tend to have mixed ethnic backgrounds, the picture is more complicated. Ethnic groups such as Native Americans, Afro-Caribbeans, and Asians often retain their racial tendency to be lactose intolerant. Ethnic groups such as Caucasians normally retain the ability to produce lactase after weaning, but even within this group many people develop lactose intolerance for one of two reasons.

- Some babies are born with a complete absence of lactase in the gut—a congenital condition called alactasia.
- In others, who initially produce lactase normally, the levels drop with increasing age. A diet rich in dairy products or a significantly reduced level of lactase can give rise to symptoms in such cases.

What are the symptoms?

Babies with alactasia develop the following symptoms immediately after they are fed with cow's milk products:

- vomiting;
- colic; and
- failure to gain weight.

People who develop lactose intolerance later in life will notice bloating, discomfort, and diarrhea occurring as the undigested lactose is broken down by the GI bacteria.

How is it diagnosed?

Lactose tolerance testing used to be the standard method used to diagnose the condition, but is rarely performed now. The patient fasts for eight hours and is then given a dose of lactose. Blood samples are then taken at regular intervals to measure blood glucose (sugar) levels. If levels do not rise, this indicates that the lactose is not being broken down. Alternatively, the patient may be administered a modified version of this test, using a dose of lactose labeled with hydrogen. In lactose intolerance, the hydrogen can then be detected on the patient's breath.

In the U.S., 25 percent of Caucasians are lactose intolerant, but 75–90 percent of Native Americans, Afro-Americans, and Asian Americans have this condition.

What are the treatment options?

The simplest solution is to exclude dairy products from the diet. Young children with alactasia require total exclusion, but adults can usually tolerate small amounts of lactose. Discovering exactly how much can be tolerated involves a process of trial and error, and the amount is likely to be different for each affected individual.

OBSTRUCTION OF THE BOWEL

Any blockage that prevents the normal passage of food through the gastrointestinal tract.

A wide variety of conditions can lead to obstruction of the bowel. Generally, the site of the blockage determines the symptoms. Obstruction of the bowel can have serious consequences, from acute pain to malnutrition.

What are the causes?

There are numerous causes of bowel obstruction, which fall into two groups: abnormal functioning of the bowel and structural problems. Functional problems can result from

- a chemical imbalance in the body, usually of calcium, which can paralyze the muscle within the intestinal wall;
- handling the bowel during surgery, which often results in the muscle tissue "going on strike," a condition known as paralytic ileus; and
- Hirschsprung's disease, where the nerves that stimulate peristalsis do not function properly (p. 148).

Without muscle activity, there is no peristalsis to keep chyme—the mixture of food and GI secretions—moving forward, and the contents of the intestines stagnate. The intestinal lining responds by secreting fluid, which combines with gas produced by bacteria to distend the gut, resulting in bloating and vomiting.

Structural problems cause actual physical blockage or interfere with GI movements. Structural bowel obstruction can be caused by many different conditions, including

- a hernia, where a loop of intestine may become trapped and twisted outside the abdominal wall, preventing the intestinal contents from passing through;
- tumors of the colon, which may completely block it;
- strictures caused by inflammation of Crohn's disease;
- twists in the intestine causing a volvulus, a kink that obstructs the passage of bowel contents;
- intussusception, a condition that affects newborns, where the intestine telescopes inside itself so that the outer layer squeezes the inner layer, blocking it;
- bands of fibrous tissue, called adhesions, that form either following surgery on the bowel or as a result of inflammation in the bowel, for example after peritonitis. Adhesions act as a belt, squeezing around the bowel and effectively blocking it; and
- atresia, where a portion of the bowel fails to develop properly, resulting in a blind end. This most commonly occurs in the duodenum or anus and is usually found in babies who are unable to feed because of the obstruction.

What are the symptoms?

Preventing the passage of food through the bowel results in four main symptoms, all of which come on suddenly. They include

- abdominal pain;
- bloating;
- vomiting; and
- total constipation.

Depending on the site of the obstruction, the patient may not even be able to pass gas.

How is it diagnosed?

The symptoms immediately indicate to a doctor that the bowel is obstructed. If a person is experiencing intermittent rather than continuous symptoms, which may happen with Crohn's strictures or with adhesions, then an upper GI barium study and small bowel follow-through (pp. 106–7), or a CT scan (p. 104), may confirm the diagnosis.

What are the treatment options?

Most cases of acute obstruction require immediate surgery (p. 130). Surgeons can remove tumors, relieve a volvulus or intussusception, and repair atresias. The success of surgery for these conditions depends on the cause. If tumors have spread before surgery is carried out, the outlook is poor. In contrast, volvulus rarely recurs because the colon is stitched to hold it in place.

Intermittent bowel obstructions caused by adhesions are usually treated by laparoscopic (keyhole) surgery in order to avoid handling the bowel, as this in itself may cause further adhesions to form. Whatever the treatment, however, adhesions often recur.

PARASITIC INFECTIONS

Intestinal infections caused by parasitic worms

Parasitic worms are now rare in developed countries but are commonly present in the intestines of people in under-developed countries with unprocessed drinking water. The worms latch onto the intestinal wall and feed on blood from the host or on the contents of the digestive tract.

What are the causes?

There are several types of parasitic worm that colonize human intestines. Tapeworms are found in countries where beef and pork are consumed, especially if the meat is not thoroughly cooked. Threadworm infection usually affects children in temperate or cold climates, such as in Northern

Europe, although it is comparatively rare. Hookworm, ascaris, and strongyloides are found mainly in underdeveloped countries with poor sanitation.

What are the symptoms?

Worm infections sometimes produce no symptoms at all, but they may cause

- abdominal pain;
- diarrhea or constipation;
- weight loss and malabsorption (in severe infections);
- anemia because of infection with hookworm; and
- severe nighttime itching around the anus, where the threadworm lays its eggs.

How is it diagnosed?

The most effective way to diagnose parasitic infections of the gastrointestinal tract is to examine the contents of the bowel (p. 100). The eggs of the infecting worm can often be identified in a stool sample; sometimes the worms themselves are visible. Blood tests may show evidence of infection, anemia or both (p. 99). An increase in the number of a specific type of white blood cell (eosinophils) may also be seen. In some cases, the doctor will carry out an endoscopic procedure such as colonoscopy (p. 115) or upper gastrointestinal endoscopy (p. 110) and is able to see the worms in the stomach or intestine.

Tapeworms grow inside the intestine and can reach lengths of up to 33 feet.

The only effective treatment is with specific antibiotics that destroy the infecting worm. In cases of threadworm, the patient's whole family will need treating. Improved food hygiene and good sanitation should avoid further infections.

PERITONITIS

Inflammation of the membrane that lines the abdominal cavity and covers the abdominal organs

Peritonitis is a very serious condition that occurs when the delicate membrane that surrounds and protects the abdominal organs becomes infected or inflamed. Infection can occur if an inflamed appendix perforates, releasing bacteria into the abdominal cavity. Peritonitis also develops if the bowel or stomach perforates, for instance because of a peptic ulcer or inflammatory bowel disease, allowing acidic stomach or intestinal contents to leak through the perforation into the abdominal cavity. Rarer causes of peritonitis include tuberculosis and abdominal tumors.

What are the symptoms?

If peritonitis is caused by perforation, the symptoms are sudden and, if untreated, may be life-threatening. They include acute severe abdominal pain and nausea, often with vomiting and loss of appetite. A universal symptom of peritonitis is "guarding," where the muscles of the abdomen contract in order to keep the contents still. Symptoms may subside on their own, but infection resulting from the leak of intestinal contents or bacteria into the abdominal cavity or bloodstream can result in septicemia and shock.

How is it diagnosed?

A person with peritonitis will be in shock, with low blood pressure and a rapid heart rate. The abdomen is rigid and very tender. The cause of the peritonitis may be diagnosed from the patient's medical history—for instance known peptic ulcers or abdominal pain suggestive of appendicitis.

What are the treatment options?

Peritonitis almost always requires surgery to repair the perforation and wash out the abdominal cavity, in order to reduce the inflammation and infection. A drain may be left in the abdomen to allow any new fluid to be drained off.

What is the outlook?

Peritonitis can be life-threatening. The onset of infection around the bowel in the abdomen usually results in septicemia—spread of infection into the bloodstream—which requires intensive care. Once the septicemia is treated, however, the patient should make a full recovery.

PEPTIC ULCER DISEASE

Inflamed ulcers in the lining of the stomach or the duodenum (the first part of the small intestine)

Peptic ulcers can occur either in the stomach or in the duodenum. Duodenal ulcers are about twice as common as gastric (stomach) ulcers; on average, about 15 percent of the population will suffer from a duodenal ulcer at some time in their lives. It was long thought that stress and anxiety contributed to the formation of peptic ulcers, but experts are now confident that this is not the case.

What are the causes?

The primary cause of peptic ulcers is the bacterium *Helicobacter pylori*, but certain drugs can also cause peptic ulcers. These include nonsteroidal antiinflammatories, such as aspirin, and steroids. This is because these drugs prevent the formation of the protective layer of mucus in the stomach or intestines, allowing acid produced in the stomach to damage the lining. Tumors of the stomach can also lead to ulceration of the lining, as can excess production of stomach acid.

What are the symptoms?

It is not usually possible to determine the location of an ulcer from its symptoms, because both gastric and duodenal ulcers can cause the following problems:

- dyspepsia, commonly called indigestion;
- abdominal pain after eating; and
- nausea or loss of appetite.

In some people, peptic ulcers are completely asymptomatic. At the other end of the scale, an ulcer may form close to a blood vessel in the stomach or duodenal wall and cause a severe gastrointestinal hemorrhage.

How is it diagnosed?

Peptic ulcers are almost always diagnosed by upper gastrointestinal endoscopy (p. 110), during which the doctor is able to directly view the stomach lining. Ulcers sometimes show up on upper GI barium study X-rays (p. 106), although this test is rarely carried out these days. Small samples of tissue—biopsies—can be taken from the ulcer to check for *Helicobacter pylori* infection and to ensure that it is not malignant.

At least 90 percent of all duodenal ulcers and 80 percent of all stomach ulcers are caused by infection by Helicobacter pylori.

What are the treatment options?

If doctors diagnose infection with *Helicobacter pylori*, a week-long prescription of two types of antibiotic (pp. 122–23), plus a third medication to reduce acid production in the stomach (p. 119), eliminates the bacteria in 90 percent of cases. This treatment cures most peptic ulcers.

If the ulcer is caused by a drug, stopping the culprit and prescribing an alternative, if necessary, allows the ulcer to heal. If it is not possible to stop the drug, the doctor will probably prescribe a strong acid inhibitor (pp. 118–19) to help the ulcer heal and to prevent it from recurring. A cancerous ulcer usually requires surgical treatment.

What is the outlook?

Once treatment has allowed a peptic ulcer to heal, and the factors that triggered its formation have been removed, the chance of a recurrence is about 1 in 100.

PYLORIC STENOSIS

Narrowing of the pylorus (the sphincter that separates the stomach from the duodenum). If the pylorus becomes too narrow, food cannot leave the stomach.

Pyloric stenosis in adults is usually the result of duodenal ulcer disease, where scarring due to the disease narrows and stiffens the pylorus. Cancers of the stomach can sometimes cause the same symptoms. In babies it is a congenital abnormality that is seen more commonly in boys than girls. The muscle that forms the sphincter is thickened, narrowing the pyloric channel.

What are the symptoms?

Pyloric stenosis in babies results in "projectile" vomiting, usually starting in the second week of life. In adults, it causes bloating, nausea, and projectile vomiting of large volumes of food and fluid. It is rare to suffer abdominal pain because of the stenosis itself, although there may be pain caused by the ulcers that caused the stenosis.

How is it diagnosed?

Doctors usually diagnose this condition using gastroscopy (upper gastrointestinal endoscopy, p. 110) or one of the many barium investigations (pp. 106–8).

What are the treatment options?

In babies, a small operation is carried out to cut through the thickened muscle; this is very effective at relieving the symptoms. In adults, the pylorus can be stretched using a balloon passed down inside an endoscope and inflated in the narrowed area (p. 124). If this is not successful, then surgery to open up the pylorus is carried out.

If a cancer is the cause of the obstruction, then surgeons can remove the affected part—a gastrectomy, for example, is the removal of the stomach—and rejoin the cut ends, which can help to relieve symptoms, although this obviously has severe implications for the patient's digestion.

What is the outlook?

In babies, surgery can permanently cure the problem. In adults, balloon stretching may not completely resolve the problem and symptoms may return, requiring surgery.

TUMORS

In Britain, tumors of the gastrointestinal tract are the most common type of cancer.

The three main types of digestive cancers are esophageal (which affects the passage leading from the throat to the stomach), gastric (which affects the stomach), and colorectal, affecting the large intestine.

What are the causes?

Smoking is a major risk factor for all types of tumor, as are poor diet and a family history of gastrointestinal cancer. More specific risk factors include long-term gastroesophageal reflux for esophageal cancer, infection with the bacterium *Helicobacter pylori*, pernicious anemia, and previous operations on the stomach for gastric cancer.

What are the symptoms?

Symptoms vary according to the type of cancer.

- Esophageal cancer causes pain and difficulty in swallowing. Initially only solid foods cause problems, but as the tumor grows, swallowing liquids also results in regurgitation. Weight loss almost always occurs.
- Tumors arising in the stomach are often very advanced before causing symptoms. Pain, feeling full after small meals, nausea, vomiting, and vomiting blood can all be caused by stomach cancer, often associated with weight loss.
- Colorectal cancers cause bowel habit changes, anemia, rectal bleeding, pain, weight loss, and bowel obstruction.

How is it diagnosed?

There are two main methods of diagnosis.

- **Endoscopy** Doctors examine the upper or lower digestive tract, depending on the site of the cancer (pp. 110–15).
- **Barium studies** The type of study depends on the type of suspected cancer: upper GI barium study for esophageal cancer (pp. 106–7), barium meal for gastric cancer (pp. 106–7), and barium enema for colorectal cancer (p. 108).

What are the treatment options?

Surgery is the main treatment option—if the cancer has not spread, it may be possible to remove it entirely by cutting out a section of the digestive tract. In patients with gastric cancer, for instance, doctors might surgically remove the stomach (a gastrectomy). Chemotherapy and radiation may be used to reduce the risk of cancer returning, and also if the cancer has spread beyond the GI system. Some cancers need special procedures—for example, inoperable esophageal cancer may require insertion of a stent, a hollow tube that keeps the esophagus open, to allow the patient to eat (p. 124). Colorectal cancer patients may need a permanent colostomy (p. 131).

What is the outlook?

Chances of full recovery tend to be poor, because most cases are not diagnosed until they are well advanced. However, the outlook for colon cancer is improving as treatment improves. If the tumor is completely removed and has not spread, then a cure is likely. Colonoscopy (p. 115) is used as a screening tool for colorectal cancer.

ULCERATIVE COLITIS

See Inflammatory bowel disease, p. 148.

LIVING WITH A COLOSTOMY

I have suffered from ulcerative colitis for many years, and after a particularly severe attack a few years ago my specialist recommended that I have a colostomy.

I was very reluctant, because I knew it would mean having a colostomy bag and I was frightened of how it would make me feel about my body. Would people be able to see it? Or worse still, smell it? Would I still be able to go swimming? And what about things like sex? Eventually, the pain of my condition forced me to reconsider and although it has taken me some time to get used to it, living with "the bag" has not been as bad as I'd imagined—few people even know I wear one.

A specialist stoma nurse taught me how to change the bag, which I have to do after every bowel movement. After the operation these were frequent, but the nurse assured me they would settle down after a few weeks, and they did—now they're like clockwork. I have to watch what I eat, because some foods create a lot of gas, which can blow up the bag! Also, my skin sometimes gets irritated by the adhesive used to attach the bags. I did have to give up swimming, but my family, and especially my husband has been very supportive, and my life is pretty well back to normal—better, because I'm not in pain.

Index

Acknowledgments

Carroll & Brown Limited would also like to thank:

Editorial assistant
Charlotte Beech

Picture researchers
Richard Soar, Sandra Schneider

Production manager
Karol Davies

Production controller
Nigel Reed

Computer management
Elisa Merino, Paul Stradling

Indexer
Kathy Croom

3-D anatomy
Mirashade/Matt Gould, Lee Harding

Illustrators
Andy Baker, Jacey, Debbie Maizels, Mikki Rain, Nick Veasey, Philip Wilson, John Woodcock

Photographers
Jules Selmes, David Murray

Photographic sources

1 SPL
6 *(left)* Biophoto Associates
(right) Prof. S. Cinti/CNRI/SPL
7 Howard Sochurek/MII/Medipics
8 *(top)* Images Colour Library
(top left) BSIP Dr. Pichard T/SPL
9 *(top)* Telegraph Colour Library
(bottom) Alfred Pasieka/SPL
10 *(top left)* GettyOne Stone
(bottom) Scott Camazine/SPL
10/11 GettyOne Stone
(background) Images Colour Library
11 *(top)* BSIP Dr Pichard T/SPL
(bottom left) GettyOne Stone
12 *(top)* Biophoto Associates/SPL
13 TEK Image/SPL
14 *(left)* Dan McCoy/Rainbow/SPL
(right) GettyOne Stone
15 Scott Camazine/SPL
16 *(bottom left)* Scott Camazine/SPL
19 *(top)* GettyOne Stone
20 *(bottom left)* Telegraph Colour Library
(bottom right) Biophoto Associates/SPL
21 *(top)* Wellcome Medical Library, Medical Arts Service, Munich
(bottom) British Technical Films/SPL
22 CNRI/SPL
24 *(right)* Telegraph Colour Library
25 *(right)* Prof. P. M. Motta/Dept. of Anatomy/University "La Sapienza," Rome/SPL
26 *(top)* Quest/SPL
(bottom) Biophoto Associates
28/9 Profs. P. M. Motta and T. Naguro/SPL
30 GettyOne Stone
32 Scott Camazine/SPL
33 Dan McCoy/Rainbow/Medipics
34 GettyOne Stone
35 *(top)* GettyOne Stone
(second from top) Pictor
38 *(top right)* Telegraph Colour Library
40 Nick Veasey/Untitled
41 Telegraph Colour Library
42 *(bottom)* GettyOne Stone
43 CNRI/SPL
45 *(top right)* Telegraph Colour Library
49 Prof. S. Cinti/CNRI/SPL
51 *(right, second from bottom)* Carl Schmidt-Luchs/SPL
(right, bottom) Telegraph Colour Library
53 Image Bank
54 *(left)* GettyOne Stone
58 Carl Schmidt-Luchs/SPL
59 GettyOne Stone
60 Telegraph Colour Library
62 *(left)* Science Pictures/Oxford Scientific Films
67 *(top)* Image Bank
(bottom centre) The Stock Market
70 Dr. Kari Lounatmaa/SPL
71 *(left)* Secchi-Lecaque/Roussel-UCLAF/CNRI/SPL
73 Gary R. Bonner
76 Nick Veasey/Untitled
83 *(left)* The Stock Market
88 *(top)* The Stock Market
89 *(top)* GettyOne Stone
90 *(left)* BSIP VEM/SPL
(right) Alfred Pasieka/SPL
91 Mehau Kulyk/SPL
94 *(bottom)* David Scharf/SPL
97 GettyOne Stone
100 *(bottom)* Profs P M Motta and F M Magliocca/SPL
101 *(left)* Jim Varney/SPL
102 *(top)* P Hawtin, University of Southampton/SPL
(bottom) John Greim/SPL
103 BSIP Dr. Pichard T/SPL
104 *(left)* BSIP Gems/Europe/SPL
(right) BSIP VEM/SPL
105 Simon Fraser/RVI, Newcastle Upon Tyne/SPL
106/7 Chris Priest/SPL
107 *(top)* Salisbury District Hospital
(bottom) SPL
108 SPL
109 *(top left, bottom right)* Kevin Harrison/Medipics
110 Dr. Klaus Schiller/SPL
111 Dr. V. S. Wong
112 *(left)* KeyMed
(right) Kevin Harrison/Medipics
113 Kevin Harrison/Medipics
115 *(top right)* BSIP VEM/SPL
(bottom) CNRI/SPL
117 *(top)* Prof. P. Motta/Dept. of Anatomy/University "La Sapienza," Rome/SPL
(bottom) Profs. P. M. Motta, K. R. Porter and P. M. Andrews/SPL
119 *(left)* Prof. J. James/SPL
(right) Manfred Kage/SPL
120 *(top)* Dr. Kari Lounatmaa/SPL
122 *(top)* Profs. P. M. Motta and F M Magliocca/SPL
(bottom) Howard Sochurek/MII/Medipics
123 *(top)* BSIP VEM/SPL
(bottom) Alfred Pasieka/SPL
124 *(left)* Kevin Harrison/Medipics
(right) Volker Steger/SPL
125 *(left)* BSIP Laurent/H American/SPL
(right) Dr. V. S. Wong
126 Antonia Reeve/SPL
127 *(left)* Mehau Kulyk/SPL
(centre) Biophoto Associates/SPL
(right) Mig/Medipics
128/9 Kevin Harrison/Medipics
130 SPL
131 Robert Harris/Medipics

619–003–2